Communications
in Computer and Information Science 65

Dominik Ślęzak   Tai-hoon Kim
Adrian Stoica   Byeong-Ho Kang (Eds.)

Control
and Automation

International Conference, CA 2009
Held as Part of the Future Generation
Information Technology Conference, CA 2009
Jeju Island, Korea, December 10-12, 2009
Proceedings

 Springer

Volume Editors

Dominik Ślęzak
University of Warsaw and Infobright Inc.,
Warsaw, Poland
E-mail: slezak@infobright.com

Tai-hoon Kim
Hannam University, 306-791 Daejeon, South Korea
E-mail: taihoonn@hnu.kr

Adrian Stoica
Jet Propulsion Laboratory/Caltech, NASA
4800 Oak Grove Drive, Pasadena, CA 91109 USA
E-mail: adrian.stoica@jpl.nasa.gov

Byeong-Ho Kang
University of Tasmania, 7001 Hobart, Australia
E-mail: bhkang@utas.edu.au

Library of Congress Control Number: 2009940004

CR Subject Classification (1998): C.3, D.4.1, F.3.2, J.7, J.2

ISSN   1867-8211
ISBN-10   3-642-10742-7 Springer Berlin Heidelberg New York
ISBN-13   978-3-642-10742-9 Springer Berlin Heidelberg New York

springer.com

© Springer-Verlag Berlin Heidelberg 2009
Printed in Germany

Typesetting: Camera-ready by author, data conversion by Scientific Publishing Services, Chennai, India
Printed on acid-free paper   SPIN: 12804368   06/3180   5 4 3 2 1 0

Foreword

As future generation information technology (FGIT) becomes specialized and fragmented, it is easy to lose sight that many topics in FGIT have common threads and, because of this, advances in one discipline may be transmitted to others. Presentation of recent results obtained in different disciplines encourages this interchange for the advancement of FGIT as a whole. Of particular interest are hybrid solutions that combine ideas taken from multiple disciplines in order to achieve something more significant than the sum of the individual parts. Through such hybrid philosophy, a new principle can be discovered, which has the propensity to propagate throughout multifaceted disciplines.

FGIT 2009 was the first mega-conference that attempted to follow the above idea of hybridization in FGIT in a form of multiple events related to particular disciplines of IT, conducted by separate scientific committees, but coordinated in order to expose the most important contributions. It included the following international conferences: Advanced Software Engineering and Its Applications (ASEA), Bio-Science and Bio-Technology (BSBT), Control and Automation (CA), Database Theory and Application (DTA), Disaster Recovery and Business Continuity (DRBC; published independently), Future Generation Communication and Networking (FGCN) that was combined with Advanced Communication and Networking (ACN), Grid and Distributed Computing (GDC), Multimedia, Computer Graphics and Broadcasting (MulGraB), Security Technology (SecTech), Signal Processing, Image Processing and Pattern Recognition (SIP), and u- and e-Service, Science and Technology (UNESST).

We acknowledge the great effort of all the Chairs and the members of advisory boards and Program Committees of the above-listed events, who selected 28% of over 1,050 submissions, following a rigorous peer-review process. Special thanks go to the following organizations supporting FGIT 2009: ECSIS, Korean Institute of Information Technology, Australian Computer Society, SERSC, Springer LNCS/CCIS, COEIA, ICC Jeju, ISEP/IPP, GECAD, PoDIT, Business Community Partnership, Brno University of Technology, KISA, K-NBTC and National Taipei University of Education.

We are very grateful to the following speakers who accepted our invitation and helped to meet the objectives of FGIT 2009: Ruay-Shiung Chang (National Dong Hwa University, Taiwan), Jack Dongarra (University of Tennessee, USA), Xiaohua (Tony) Hu (Drexel University, USA), Irwin King (Chinese University of Hong Kong, Hong Kong), Carlos Ramos (Polytechnic of Porto, Portugal), Timothy K. Shih (Asia University, Taiwan), Peter M.A. Sloot (University of Amsterdam, The Netherlands), Kyu-Young Whang (KAIST, South Korea), and Stephen S. Yau (Arizona State University, USA).

We would also like to thank Rosslin John Robles, Maricel O. Balitanas, Farkhod Alisherov Alisherovish, and Feruza Sattarova Yusfovna – graduate students of Hannam University who helped in editing the FGIT 2009 material with a great passion.

October 2009

Young-hoon Lee
Tai-hoon Kim
Wai-chi Fang
Dominik Ślęzak

Preface

We would like to welcome you to the 2009 International Conference on Control and Automation (CA 2009), which was organized as a part of the 2009 International Mega-Conference on Future Generation Information Technology (FGIT 2009), held during December 10-12, 2009, at the International Convention Center Jeju, Jeju Island, South Korea.

CA 2009 focused on various aspects of advances in control and automation. It provided a chance for academic and industry professionals to discuss recent progress in the related areas. We expect that the conference and its publications will be a trigger for further related research and technology improvements in this important subject.

We would like to acknowledge the great effort of all the Chairs and members of the Program Committee. Out of 40 submissions to CA 2009, we accepted 12 papers to be included in the proceedings and presented during the conference. This gives a roughly 30% acceptance ratio. Two of the papers accepted for CA 2009 were published in the special FGIT 2009 volume, LNCS 5899, by Springer. The remaining 10 accepted papers can be found in this CCIS volume.

We would like to express our gratitude to all of the authors of submitted papers and to all of the attendees, for their contributions and participation. We believe in the need for continuing this undertaking in the future.

Once more, we would like to thank all the organizations and individuals who supported FGIT 2009 as a whole and, in particular, helped in the success of CA 2009.

October 2009

Dominik Ślęzak
Tai-hoon Kim
Adrian Stoica
Byeong-Ho Kang

Organization

Organizing Committee

General Chair   Adrian Stoica (NASA Jet Propulsion Laboratory, USA)

Program Chairs   Byeong-Ho Kang (University of Tasmania, Australia)
                       Tai-hoon Kim (Hannam University, Korea)

Program Committee

Albert Cheng
Alessandro Casavola
Barry Lennox
Bernard Grabot
Choonsuk Oh
Christian Schmid
Chun-Yi Su
DaeEun Kim
DongWon Kim
Feng-Li Lian
Guang-Ren Duan

Guoping Liu
Gwi-Tae Park
Hideyuki Sawada
Hojjat Adeli
Hong Wang
Jong H. Park
Jong-Wook Kim
Jurek Sasiadek
Jus Kocijan
Kwan-Ho You
Kwon Soon Lee

Makoto Itoh
Manuel Haro Casado
Mitsuji Sampei
Myotaeg Lim
Peter Simon Sapaty
Pierre Borne
Pieter J. Mosterman
S.C. Kim
Thomas Parisini
Zhong-Ping Jiang
Zuwairie Ib

Table of Contents

Necessary and Sufficient Conditions on Quadratic Stablizability for Polytopic Uncertain Continuous-Time Switched Linear Systems

Naohisa Otsuka[1] and Takuya Soga[2]

[1] Division of Science, School of Science and Engineering,
Tokyo Denki University,
Hatayama-Machi, Hiki-Gun, Saitama 350-0394, Japan
otsuka@mail.dendai.ac.jp
[2] Information Sciences, Graduate School of Science and Engineering,
Tokyo Denki University,
Hatayama-Machi, Hiki-Gun, Saitama 350-0394, Japan

Abstract. In this paper we study quadratic stabilizability via state feedback for continuous-time switched linear systems whose subsystem's matrices are represented as a polytopic linear combination of vertex matrices. Necessary and sufficient conditions for polytopic uncertain continuous - time switched linear systems composed of two subsystems to be quadratically stabilizable via state feedback are proved.

Keywords: Switched Linear Systems, Quadratic Stabilizability, Polytopic Uncertain Systems.

1 Introduction

The so-called switched system is represented as the family of subsystems with switching rule which concerns with various environmental factors and different controllers. For such switched systems it is important to investigate the stability and stabilizability as fundamental problems. Until now many results on stability and stabilizability problems for various types of switched systems have been studied (e.g., [1],[3]-[6],[10],[12]-[14]). Further, many interesting results for various problems of switched systems have been written in some books (e.g., [7],[8],[11]).

On the other hand, from the practical viewpoint, it is important to investigate switched systems which contain uncertain parameters. Recently, Lin and Antsaklis[9] and Zhai, Lin and Antsaklis[15] investigated the stability and stabilizability problems for both continuous-time and discrete-time uncertain switched linear systems. Especially, the paper[15] investigated the quadratic stabilizability problem via state feedback for polytopic uncertain switched linear systems, and sufficient conditions for the switched systems which composed of two subsystems to be quadratically stabilizable were given. However, necessary

D. Ślęzak et al. (Eds.): CA 2009, CCIS 65, pp. 1–7, 2009.
© Springer-Verlag Berlin Heidelberg 2009

and sufficient conditions for the switched systems which composed of two sub-systems have not been given even with two subsystems for both continuous-time and discrete-time cases.

In this paper we study quadratic stabilizability via state feedback for contin-uous - time switched linear systems whose subsystem's matrices are represented as a polytopic linear combination of vertex matrices. In Section 2 necessary and sufficient conditions for polytopic uncertain switched linear systems to be quadratically stabilizable via state-feedback are proved for the case of two sub-systems by using the so-called "S-procedure Lemma". The obtained results are extensions of the results of Feron[4]. Section 3 gives some concluding remarks.

2   Polytopic Uncertain Switched Systems

Consider the following continuous-time switched linear system

$$\Sigma_s : \dot{x}(t) = A_{\sigma(x,t)}x(t), \ x(0) = x_0$$

where $x(t) \in \Re^n$ is the state vector, $\sigma(x,t) : \Re^n \times \Re^+ \rightarrow \{1, 2, \cdots, L\}$ is a switched rule which depends on the state x and time t, and $\Re^+$ is the set of non-negative real numbers.

Then, the above switched system Σ_s is composed of the family of continuous-time subsystems

$$\Sigma_i : \dot{x}(t) = A_i x(t), \ i = 1, 2, \cdots, L,$$

where $L \ (\geq 2)$ is the number of subsystems. The architecture of the switched system Σ_s is shown as Fig 1.

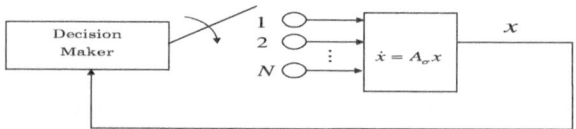

Fig. 1. Architecture of switched system Σ_s

In this section we consider the case that all subsystems have polytopic uncer-tainties described as

$$A_i = \sum_{j=1}^{N_i} \mu_{i,j} A_{i,j}, \ \sum_{j=1}^{N_i} \mu_{i,j} = 1, \ (i = 1, 2, \cdots, L), \quad (1)$$

where $A_{i,j} \ (j = 1, 2, \cdots, N_i)$ are the vertex matrices of the polytopic matrix A_i, N_i is the number of the vertex matrices $A_{i,j}$ and $\mu_{i,j} \ (j = 1, 2, \cdots, N_i)$ are polytopic uncertain parameters for each $i = 1, \cdots, L$.

Now, we give the definition of quadratic stabilizability via state feedback for the switched system Σ_s.

Definition 1. The switched linear system Σ_s is said to be quadratically stabilizable via state feedback if there exist a Lyapunov function of the form $V(x) = x^T P x$, a positive number $\epsilon\ (>0)$ and a switched rule $\sigma(x,t)$ such that

$$\frac{d}{dt}V(x) < -\epsilon x^T x$$

for all trajectory x of the system Σ_s.

When we investigate the quadratic stabilizability problem of switched systems, the following assumption is given. Because, if there is a stable subsystem in the family of subsystems, we can always activate the stable subsystem and therefore the problem becomes trivial one.

Assumption 2. Assume that all subsystems $\Sigma_i\ (i = 1, 2, \cdots, L)$ are asymptotically unstable, which imply there does not exist positive definite matrices $P_i > 0\ (i = 1, \cdots, L)$ such that

$$A_{i,j}^T P_i + P_i A_{i,j} < 0 \quad (j = 1, 2, \cdots, N_i).$$

Before giving the main theorem, we need the following lemma which is known as the so-called "S-procedure".

Lemma 3. ([2],[7]) Let T_0 and $T_1 \in \Re^{n \times n}$ be two symmetric matrices. Consider the following two conditions:

$$x^T T_0 x > 0 \text{ whenever } x^T T_1 x \geq 0 \text{ and } x \neq 0 \tag{2}$$

and

$$\exists \tau_1 \geq 0 \text{ such that } T_0 - \tau_1 T_1 > 0. \tag{3}$$

Then, condition (3) always implies condition (2). If there is some $x_0 \neq 0$ such that $x_0^T T_1 x_0 > 0$, then condition (2) implies condition (3).

When the number of subsystems is two (i.e., $L = 2$), the result of sufficiency in the theorem shown by Feron[4] was extended to the polytopic uncertain switched linear system by Zhai et al.[15]. However, the necessity has not been proved.

The following theorem gives necessary and sufficient conditions for the polytopic uncertain switched linear system to be quadratically stabilizable via state feedback.

Theorem 4. Suppose that $L = 2$ and Assumption 2 holds. The polytopic uncertain switched linear system Σ_s with (1) is quadratically stabilizable via state feedback if and only if there exist $\lambda_{k,j} \in [0,1]\ (k = 1, \cdots, N_1;\ j = 1, \cdots, N_2)$ such that $\lambda_{k,j} A_{1k} + (1 - \lambda_{k,j}) A_{2j}$ are simultaneously asymptotically stable, that is, there exists a positive definite matrix $P > 0$ such that

$$[\lambda_{k,j} A_{1k} + (1 - \lambda_{k,j}) A_{2j}]^T P + P[\lambda_{k,j} A_{1k} + (1 - \lambda_{k,j}) A_{2j}] < 0,$$

equivalently

$$\lambda_{k,j}(A_{1k}^{\mathrm{T}}P + PA_{1k}) + (1 - \lambda_{k,j})\,(A_{2j}^{\mathrm{T}}P + PA_{2j}) < 0.$$

Proof. (Sufficiency) The proof of sufficiency was given by Zhai et al.[15].

(Necessity) Suppose that the polytopic uncertain switched linear system Σ_s with (1) is quadratically stabilizable via state feedback. Then, there exist an $\epsilon > 0$, $P > 0$ and a switched rule $\sigma(x, t)$ such that

$$
\begin{aligned}
\frac{d}{dt}\,V\{x(t)\} &= \frac{d}{dt}\,\{x^{\mathrm{T}}(t)Px(t)\} \\
&= \dot{x}^{\mathrm{T}}(t)Px(t) + x^{\mathrm{T}}(t)P\dot{x}(t) \\
&= \{A_{\sigma(x,t)}x(t)\}^{\mathrm{T}}Px(t) + x^{\mathrm{T}}(t)P\{A_{\sigma(x,t)}x(t)\} \\
&= x^{\mathrm{T}}(t)\{A_{\sigma(x,t)}^{\mathrm{T}}P + PA_{\sigma(x,t)}\}x(t) \\
&< -\epsilon x^{\mathrm{T}}(t)x(t)
\end{aligned}
\tag{4}
$$

for all trajectory $x(t)$ of the system Σ_s. Then, it follows from (4) that we have either

$$x^{\mathrm{T}}(t)(A_1^{\mathrm{T}}P + PA_1)x(t) < -\epsilon x^{\mathrm{T}}(t)x(t) \tag{5}$$

or

$$x^{\mathrm{T}}(t)(A_2^{\mathrm{T}}P + PA_2)x(t) < -\epsilon x^{\mathrm{T}}(t)x(t) \tag{6}$$

for arbitrary state $x(t)$.

Since A_1 and A_2 are asymptotically unstable from Assumption 2 and (5) or (6), there exists an $x(t)$ satisfying

$$
\begin{cases}
x^{\mathrm{T}}(t)(A_1^{\mathrm{T}}P + PA_1)x(t) < -\epsilon x^{\mathrm{T}}(t)x(t) \\
\text{whenever } x^{\mathrm{T}}(t)(A_2^{\mathrm{T}}P + PA_2)x(t) \geq -\epsilon x^{\mathrm{T}}(t)x(t)
\end{cases}
\tag{7}
$$

or

$$
\begin{cases}
x^{\mathrm{T}}(t)(A_2^{\mathrm{T}}P + PA_2)x(t) < -\epsilon x^{\mathrm{T}}(t)x(t) \\
\text{whenever } x^{\mathrm{T}}(t)(A_1^{\mathrm{T}}P + PA_1)x(t) \geq -\epsilon x^{\mathrm{T}}(t)x(t).
\end{cases}
\tag{8}
$$

Suppose that there exists an $x(t)$ satisfying (7). Then,

$$
\begin{cases}
-x^{\mathrm{T}}(t)(A_1^{\mathrm{T}}P + PA_1 + \epsilon I)x(t) > 0 \\
\text{whenever } x^{\mathrm{T}}(t)(A_2^{\mathrm{T}}P + PA_2 + \epsilon I)x(t) \geq 0.
\end{cases}
\tag{9}
$$

If we consider (1) in (9), we have

$$
\begin{cases}
-x^{\mathrm{T}}(t)\left\{\left(\displaystyle\sum_{j=1}^{N_1}\mu_{1,j}A_{1,j}\right)^{\mathrm{T}}P+P\left(\displaystyle\sum_{j=1}^{N_1}\mu_{1,j}A_{1,j}\right)+\epsilon I\right\}x(t)>0\\[4ex]
\text{whenever}\\[1ex]
x^{\mathrm{T}}(t)\left\{\left(\displaystyle\sum_{j=1}^{N_2}\mu_{2,j}A_{2,j}\right)^{\mathrm{T}}P+P\left(\displaystyle\sum_{j=1}^{N_2}\mu_{2,j}A_{2,j}\right)+\epsilon I\right\}x(t)\geq0,
\end{cases}\tag{10}
$$

where $\displaystyle\sum_{j=1}^{N_i}\mu_{1,j}=1$ $(i=1,2)$.

Now, let choose elements $\mu_{1,k}$ and $\mu_{2,j}$ $(k=1,\cdots,N_1;\ j=1,\cdots,N_2)$ in (10) as

$$
\mu_{1,k}=1\ \ (\text{ i.e. }\mu_{1,\ell}=0,\ \ell\neq k),
$$
$$
\mu_{2,j}=1\ \ (\text{ i.e. }\mu_{2,p}=0,\ p\neq j)
$$

for $k=1,\cdots,N_1$ and $j=1,\cdots,N_2$. Then, we have

$$
\begin{cases}
-x^{\mathrm{T}}(t)(A_{1,k}^{\mathrm{T}}P+PA_{1,k}+\epsilon I)x(t)>0\\
\text{whenever }\ x^{\mathrm{T}}(t)(A_{2,j}^{\mathrm{T}}P+PA_{2,j}+\epsilon I)x(t)\geq0.
\end{cases}\tag{11}
$$

Define the following two symmetric matrices in (11) as

$$
T_0:=-(A_{1,k}^{\mathrm{T}}P+PA_{1,k}+\epsilon I),
$$
$$
T_1:=A_{2,j}^{\mathrm{T}}P+PA_{2,j}+\epsilon I.
$$

Now, there exists an x_0 such that $x_0^{\mathrm{T}}T_1x_0>0$. In fact, if $x_0^{\mathrm{T}}T_1x_0\leq0$, that is,

$$
x_0^{\mathrm{T}}(A_{2,j}^{\mathrm{T}}P+PA_{2,j})x_0\leq-\epsilon x_0^{\mathrm{T}}x_0
$$

for all elements x_0 which contradicts Assumption 2.

Hence, it follows from Lemma 3 that there exists a $\tau_{11}\geq0$ such that $T_0-\tau_{k,j}T_1>0$, that is

$$
(A_{1,k}^{\mathrm{T}}P+PA_{1,k}+\epsilon I)+\tau_{k,j}(A_{2,j}^{\mathrm{T}}P+PA_{2,j}+\epsilon I)<0
$$

$$
\Leftrightarrow (A_{1,k}^{\mathrm{T}}P+PA_{1,k})+\tau_{k,j}(A_{2,j}^{\mathrm{T}}P+PA_{2,j})<-(1+\tau_{k,j})\epsilon I
$$

$$
\Leftrightarrow \frac{1}{1+\tau_{k,j}}(A_{1,k}^{\mathrm{T}}P+PA_{1,k})+\frac{\tau_{k,j}}{1+\tau_{k,j}}(A_{2,j}^{\mathrm{T}}P+PA_{2,j})<-\epsilon I.
$$

Define $\lambda_{k,j}:=\dfrac{1}{1+\tau_{k,j}}\in[0,1]$. Then, we get

$$
[\lambda_{k,j}A_{1,k}+(1-\lambda_{k,j})A_{2,j}]^{\mathrm{T}}P+P[\lambda_{k,j}A_{2,j}+(1-\lambda_{k,j})A_{2,j}]<-\epsilon I,\tag{12}
$$

for $k = 1, \cdots, N_1$ and $j = 1, \cdots, N_2$. Similarly, if there exists an $x(t)$ satisfying (8), we can get the inequality (12) in the same manner. This completes the proof of necessity.   □

If we consider no uncertainties in the switched system Σ_s, the following result which was given by Feron[4] can be obtained as a special case from Theorem 4.

Corollary 5. [4] Suppose that $L = 2$ and Assumption 2 is satisfied. Further, suppose that subsystems A_i $(i = 1, 2)$ do not contain uncertain parameters, that is, $A_i = A_{i,1}$ $(i = 1, 2)$. Then, the switched linear system Σ_s is quadratically stabilizable via state feedback if and only if there exist a $\lambda \in [0, 1]$ such that $\lambda A_1 + (1 - \lambda)A_2$ is asymptotically stable, that is, there exist a positive definite matrix $P > 0$ such that

$$[\lambda A_1 + (1 - \lambda)A_2]^\mathrm{T} P + P[\lambda A_1 + (1 - \lambda)A_2] < 0,$$

equivalently

$$\lambda(A_1^\mathrm{T} P + P A_1) + (1 - \lambda)\,(A_2^\mathrm{T} P + P A_2) < 0.$$

3   Concluding Remarks

In this paper, we investigated quadratic stabilizability problem via state feedback for polytopic uncertain continuous-time switched linear system in the sense that subsystem's matrices are represented as a polytope of vertex matrices. Necessary and sufficient conditions for polytopic uncertain switched linear systems composed of two subsystems to be quadratically stabilizable via state feedback were proved by using the results of Zhai et al.[15] and S-procedure Lemma. The obtained results are extensions of the results of Feron[4].

   As future studies, it remains to investigate necessary and sufficient conditions for the polytopic uncertain discrete-time switched linear systems to be quadratically stabilizable via state feedback.

References

1. Akar, M., Paul, A., Safonov, M.G., Mitra, U.: Condtions on the stability of a class of second-order switched systems. IEEE Transactions on Automatic Control 51(2), 338–340 (2006)
2. Boyd, S., Ghaoui, L.E., Feron, E., Balakrishnan, V.: Linear matrix inequalities in system and control theory. SIAM (1994)
3. Branicky, M.S.: Stability of switched and hybrid systems. In: Proceedings of the 33rd IEEE Conference on Decision and Control, pp. 3498–3503 (1994)
4. Feron, E.: Quadratic stablizibility of switched systems via state and output feedback., MIT Technical report CICSP-468, pp.1-13 (1996)
5. Gurvits, L., Shorten, R., Mason, O.: On the stability of switched positive linear systems. IEEE Transactions on Automatic Control 52(6), 1099–1103 (2007)
6. Liberson, D.: Basic problems in stability and design of switched systems. IEEE Control Systems Magazine 19, 59–70 (1999)

7. Liberson, D.: Switching in systems and control, Systems & Control: Foundation & Applications, Birkhäuser (2003)
8. Liu, D., Antsaklis, P.J. (eds.): Stabilty and control of dynamical systems with applications, A Tribute to Anthony N.Michel, Birkhäuser (2004)
9. Lin, H., Antsaklis, P.J.: Switching stabilizability for continuous-time uncertain switched linear systems. IEEE Transactions on Automatic Control 52(4), 633–646 (2007)
10. Mason, O., Shorten, R.: On linear copositive Lyapunov functions and the stability of switched positive linear systems. IEEE Transactions on Automatic Control 52(7), 1346–1349 (2007)
11. Sun, Z.D., Ge, S.S.: Switched linear systems control and design. Springer, Heidelberg (2004)
12. Wicks, M., Peleties, P., De Carlo, R.: Construction of piecewise Lyapunov functions for stablizing switched systems. In: Proceedings of IEEE Conference on Dicision and Control, pp. 3492–3497 (1994)
13. Wicks, M., Peleties, P., De Carlo, R.: Switched controller synthesis for the quadratic stabilization of a pair of unstable linear systems. European Journal of Control 4(2), 140–147 (1998)
14. Zhai, G.: Quadratic stablizibility of discrete-time switched systems via state and output feedback. In: Proceedings of the 40th IEEE Conference on Decision and Control, pp. 2165–2166 (2001)
15. Zhai, G., Lin, H., Antsaklis, P.J.: Quadratic stablizibility of switched systems with polytopic uncertainties. International Journal of Control 76(7), 747–753 (2003)

Identification and Robust PID Current Control of Industrial Stone Cutting Machine Using Quantitative Feedback Theory

Omid Safarzadeh

Shahid Beheshti university, Tehran, Iran
O.Safarzadeh@mail.sbu.ac.ir

Abstract. In this paper, a robust PID current control system for the stone cutting machine using the Quantitative Feedback Theory (QFT) method is presented. The stone cutting machine is a major component of stone cutting factory. To design a robust QFT controller a control oriented linear models for the stone cutting machine are identified using experimental data. The wide range uncertainties in model parameters are caused by sharpness saw blade, and different kind of stone cutting. The proposed PID controller must satisfy the disturbance rejection, robust tracking, and robust stability specifications. According to simulation results, the designed controller will ensure all the designers closed loop specifications, and the effectiveness of the robust QFT controller.

Keywords: robust control, stone cutting machine, QFT.

1 Introduction

Stone cutting machine is a major component of stone cutting factories which cut raw blocks of stone. A stone cutting machine is shown in Fig. 1. The system is highly uncertain which caused by sharpness saw blade, and different kind of stone cutting. In order to control the system, an accurate model is identified using experimental data. The essential features of the system are captured by the derived model such as the system delay. Thereafter, based on the QFT technique, a robust PID current controller for the system is designed. The resulting controller is applied to the model to show the effectiveness of the proposed controller. In this paper, the powerful systematic and transparent robust design based on the Quantitative Feedback Theory (QFT) is applied to the water level control of the horizontal steam generator. QFT analyzes the transfer function in the complex plane which includes magnitude and phase information. The basic ideas of QFT efficient robust control design was introduced by Horowitz in the early 1960s. QFT is a frequency domain technique which has been successfully implemented in many different systems such as uncertain LTI, nonlinear, non minimum phase, and unstable ones [1].

D. Ślęzak et al. (Eds.): CA 2009, CCIS 65, pp. 8–14, 2009.
© Springer-Verlag Berlin Heidelberg 2009

Fig. 1. Stone cutting machine

This technique emphasizes the fact that feedback is only necessary because of uncertainty (structured or unstructured), thus the amount of feedback should be directly related to the extent of plant uncertainty and unknown external disturbances. The main objective of QFT is for minimizing the cost of feedback, as measured by the amount of controller bandwidth [1]. Fairly simplicity and easily implementation of QFT method are some of its advantageous in control design.

The rest of this paper is organized as follows: In section 1, the model for the stone cutting machine is identified. In section 2, robust PID controller design using Quantitative Feedback theory is given. In section 3, the designed PID controller is applied to the model and the simulation results are given, and finally, concluding results are given in section 4.

2 Main Motor Current System Identification

Knowledge of main motor current of stone cutting machine is of central importance for designing a controller. The stone cutting machine is a single input-single output (SISO) system with some disturbances. The manipulated input is the forward speed of the saw blade and the output is main motor current. To identify a model a step-like input is applied to the forward movement speed of saw blade and the output is caught by data acquisition system which consist a current transformer (CT), a transducer that converts the AC current to DC voltage, and a PC oscilloscope card to watch and store the experimental data (Fig. 2). Although step inputs are not enough for identification [2], by using these data and prediction error method [3], a model is achieved which illustrated as follow:

$$I(s) = K \frac{\exp(-T_d s)}{\left(\omega^2 s^2 + 2\zeta\omega s + 1\right)} V(s) \tag{1}$$

Where the parameters are given as follow:

$K \in [20, 50]$, $\omega \in [0.003, 0.005]$, $\zeta \in [26.53, 30.54]$, $T_d = 0.13$

Fig. 2. The acquisition system

As mentioned before, the uncertainty in model parameters may caused by sharpness saw blade, and different kind of stone cutting. To compare the derived model and the real system, some step-like signals are applied to the real system and the identified model. The experiments and model results are shown in Fig. 3. As it can be observed, there is a good agreement between the model data and the experimental data.

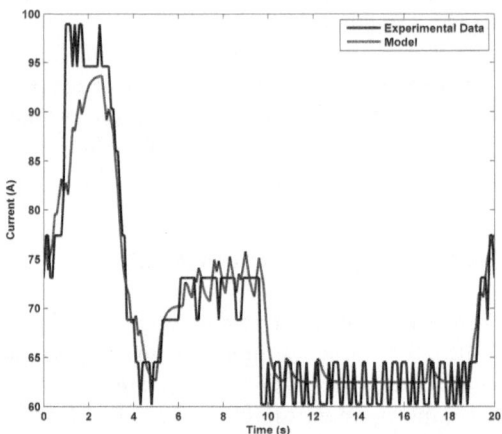

Fig. 3. Outputs of the real system and model

3   QFT Controller Design

QFT is a robust control design method, with special emphasis on the use feedback for achieving good performance against uncertainty and disturbances. The control structure is illustrated in Fig. 4 which $P(s)$ is the transfer function of the stone cutting machine, $G(s)$ is the cascade compensator; $F(s)$ is an input filter transfer function, and $D(s)$ is the output disturbance. The limitation of control voltage input of the inverter is applied by a saturation block, the controller and pre-filter will design by the powerful systematic and transparent robust design method.

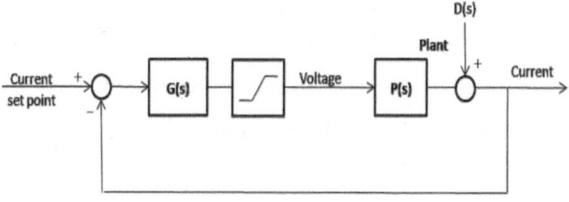

Fig. 4. QFT control structure

The design process is summarized as follows:

3.1 Template Generation

Template is a region in the Nichols or Nyquist chart that represent the frequency response of the uncertain plant at a particularly frequency. The templates of the system are shown in Fig. 5. Design frequencies are chosen based on the designers experience and some guidance. The design frequencies are usually chosen between $0 < \omega < \omega_h$ which ω_h is the frequency that the magnitude of the upper tracking performance specification is -12dB [1].

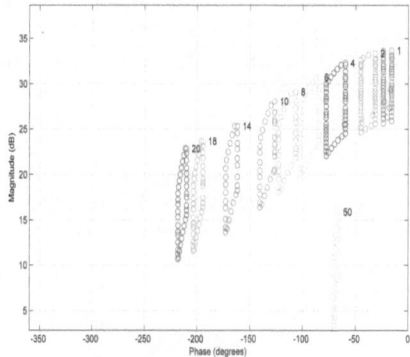

Fig. 5. Plant templates

3.2 Closed Loop Performance Specification

In QFT, the design specifications should be specified in the frequency domain. These specifications define bounds in the Nichols chart that must be satisfied by shaping of the nominal open loop transfer function. These specifications are defined as follows:

3.2.1 Robust Stability
To ensure the stability of the closed loop system for all identified plants, the following constraint is applied on the peak magnitude of the closed loop frequency response:

$$\left|\frac{\mathbf{L}}{1+\mathbf{L}}\right| \le M_L \tag{2}$$

Where the open loop transfer function with uncertainly is $\mathbf{L}$, M_L is constant over the frequency and corresponds to the minimum gain and phase margin (PM) and can be derived from the following equation [1]:

$$PM = 180° - 2\cos^{-1}(0.5/M_L) \tag{3}$$

According to the practical consideration, the minimum phase margin that we should achieve is 45°, $M_L = 1.3$.

3.2.2 Sensitivity Reduction
We used sensitivity reduction as H_∞ method for robust tracking purpose. The following constraint used for robust tracking.

$$\left|\frac{F(j\omega)}{1+PG(j\omega)}\right| \le \frac{1}{W_s(j\omega)} \tag{4}$$

$$W_s(j\omega) = \frac{s^2 + 10.8s + 0.36}{(s^2 + 10.8s + 36)(0.01s + 1)} \tag{5}$$

After computing these bounds, the composite bound should be specified this is used to synthesize the open loop transfer function. The composite bound is composed those bounds in the way that have the largest magnitude values over the phase rang. The composite bound is shown in fig. 6.

3.3 Loop Shaping

The key step in the QFT controller design is shaping the nominal open loop transfer function to satisfy the composite bound at design frequencies. There are some

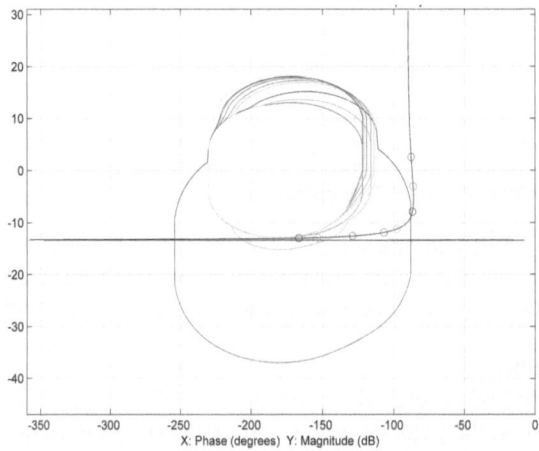

Fig. 6. The composite bound and shaped open loop transfer function

guidelines to shape the open loop transfer function which is given in [1]. After some trial and error, we have achieved an appropriate design. Fig. 6 shows the shaped open loop transfer function.

As illustrated, the open loop transfer function fulfills the composite bound, and then the controller is indicated in (6).

$$0.0316 + 0.0033s + \frac{0.066}{s} \qquad (6)$$

4   Simulation Results

The desired closed loop performance is characterized by minimum overshoot for avoiding the unplanned trip of the stone cutting machine. The output disturbances should also be rejected. The control signal, which is the inverter input voltage, should have a smooth change for practical implementation considerations.

For verifying the reference tracking QFT controller a step-like set point is applied, it is assumed the motor already is started and the no-load current is about 56 A. The response of the system is illustrated in fig. 7. The associated control signal (inverter input voltage) is shown in fig. 8.

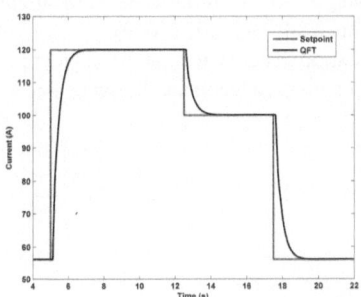

Fig. 7. The step response of the stone cutting machine current

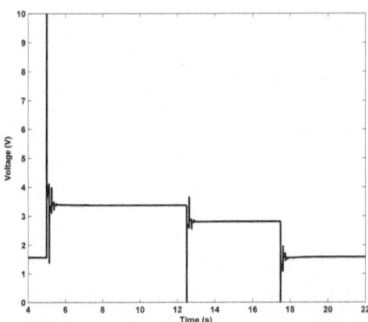

Fig. 8. The control signal to response to the associated set-point

The proposed controller has acceptable overshoot, and the settling time. The control signal is smooth and its amplitude is less than the maximum inverter input voltage. These responses clearly indicate that the designed QFT controller satisfies all the closed loop design specifications in the face of plant parameter uncertainties and output disturbances.

5 Conclusion

Stone cutting machine is a major component of stone cutting factories which cut raw blocks of stone. The system is highly uncertain which caused by sharpness saw blade, and different kind of stone cutting. In this paper, a fairly accurate model for the stone cutting machine is identified, and a robust PID controller for stone cutting machine is designed. Applying the designed controller to the model shows the effectiveness of the proposed controller.

References

1. Houpis, C.H., Rasmussen, S.J., García-Sanz, M.: Quantitative Feedback Theory, Fundamentals and Applications, Florida (2006)
2. Sontag, E.D., Yuan, W., Megretski, A.: Input Classes for Identifiability of Bilinear Systems. IEEE Trans. Automat. Control 54, 195–207 (2009)
3. Ljung, L.: System Identification: Theory for the User, New Jersey (1987)
4. Yaniv, O., Horowitz, L.: Quantitative feedback design of linear and nonlinear control systems, London (1999)

Dynamic Simulation and Synthesis Technique for Complex Control Systems

Armen Bagdasaryan[1] and Tai-hoon Kim[2]

[1] Institution of the Russian Academy of Sciences,
V. A. Trapeznikov Institute for Control Sciences of RAS,
65 Profsoyuznaya, 117997 Moscow, Russia
abagdasari@hotmail.com
[2] Dept. of Multimedia, Hannam University
133 Ojeong-dong, Daedeok-gu, Daejeon, Korea
taihoonn@hnu.kr

Abstract. A method for dynamic model synthesis and simulation of complex hierarchical control systems is developed. The method provides integration of large data sets, monitoring data and expert knowledge with the process of simulation, analysis and prediction of system state dynamics, based on the control scenarios methodology. The proposed technique is based on object-oriented approach and is powerful for information rich environments.

Keywords: Control Systems, Complexity, Hierarchy, Large-Scale Objects, Models, State Dynamics, Control Scenarios, Simulation.

1 Introduction

One of the main characteristic of modern complex control systems is that it is impossible to uniquely and adequately describe these systems, using classical mathematical methods. Classical mathematical models and approaches are applicable just for a few problem domains, which are static and comprehensible, and have most general properties. But in complex dynamic environments, with an increase of complexity, problem domains become dynamic, requiring for dynamic solutions that will be able to adapt to the changes in the problem domain [1].

Basic reasons that make it difficult for complex control systems to be described by formalized/analytical methods are the following ones:

- Information *incompleteness* on the state and the behavior of system;
- Presence of a *human* (observer, problem domain expert, etc.) as an intelligent subsystem that forms requirements and makes decisions in complex systems;
- *Uncertainty* (inconsistency, antagonism) and multiplicity of the purposes of a complex system, which can not be described in a precise formulation;
- *Restrictions* imposed on the purposes (controls, behavior, final results) externally and/or internally in relation to system are often unknown;
- *Weak structuredness*, uniqueness, combination of individual behaviors with collective ones are the intrinsic features of complex systems.

D. Ślęzak et al. (Eds.): CA 2009, CCIS 65, pp. 15–27, 2009.

Large-scale control systems typically possess a hierarchical architecture in order to manage complexity. Higher levels of the hierarchy utilize coarser models of the system, resulting from aggregating the detailed lower level models. In this layered control paradigm, the notion of hierarchical consistency is important, as it ensures the implementation of high-level objectives by the lower level system [6]. Large-scale systems are systems of very high complexity. Complexity is typically reduced by imposing a hierarchical structure on the system architecture [7]. Hierarchical structures for discrete event systems have been considered in multiple works [8], [9], [10], [2]. In such a structure, systems of higher functionality reside at higher levels of the hierarchy and are therefore unaware of unnecessary lower-level details. One of the main challenges in hierarchical systems is the extraction of a hierarchy of models at various levels of abstraction which are compatible with the functionality and objectives of each layer. The notions of abstraction or aggregation refer to grouping the system states or control objects into equivalence classes [4].

For modeling and analysis of complex control systems in the presence of principally non-formalizable problems and impossibility of strict mathematical formulation of problems, expert knowledge and information databases are used. Construction of models of complex systems is accompanied by extensive use of expert knowledge and information about the system stored in data- and information systems. This knowledge should be integrated in a unified way. Qualitative character of the most of parameters of complex systems results in knowledge fuzziness and uncertainty and, as a consequence, in problem of its formalization.

Under such conditions, the key problem of complex systems and control theory consists in the development of methods of qualitative analysis of the dynamics of such systems and in the construction of efficient control techniques. Besides this, one of the main and actual problems in the theory of complex systems and control sciences is a solution of "ill-posed, weakly-structured and weakly-formalized complex problems" associated with complex technical, organizational, social, economic, and many other objects, and with the perspectives of their evolution. Since the analysis and efficient control are impossible without a formal model of a system, the technologies for building (synthesis of) the models of complex systems have to be used.

The search for the solution of the above problems had led to the development of specialized simulation [3] and formalization [5] methods. The methods enables one to study the behavior of complex system, using computer modeling. Unlike the "traditional" methods of computer modeling, system dynamics does not require the construction of mathematical model of object under study in traditional form. It provides a tool for modeling the system elements and relations between them in the form of analytical descriptions realized in computer. Formal languages describing the change processes of modelled object are the most important constituents of system dynamics. One of these is a system diagrams language. System diagram is a tool for formalization of the model of object under study. However, construction of system diagrams in case of complex systems becomes a difficult problem, and synthesis of the object's dynamical model, acceptable for practical use, may take a very long period of time. So, the search for the ways for formalization and automation of the modeling and control processes attracts much attention.

The aim of this work is to propose a technique for construction of dynamic models of complex multi-parameter control systems consisting of a sufficiently large number of hierarchically structured large-scale objects, and also to propose an adequate state dynamics simulation tool. The main idea of the proposed method is twofold: (1) to study not the precise values of parameters describing the state of each object (trajectory of states), but *classes of objects* in framework of hierarchical structure of a set of objects, (2) to define states not as a combination of values of parameters or their ranges but as a combination of *parallel trajectories* of parameter changes at a time interval [11,12]. Classes of objects are determined by *polymorphic parameters* that characterize the objects at different levels of hierarchy. Polymorphic parameters in hierarchical models enable one to turn from control at the level of objects to control at the level of classes of objects, and also from individual models to integral models of arbitrary level of generalization. Using this approach, the problem of modeling of complex large-scale control system can be reduced to the analysis and interpretation of long-term dynamics of polymorphic parameters. Object-oriented approach provides the integrative representation of control and problem domains. Such an integral description of objects and definition of state allows us to cope with the problem of dimension, to enhance the efficiency of control decisions, and to cope with a large amount of information about a system. To formalize the problem and system dynamics, a hierarchical graph state diagrams technique [12] is used.

2   Conceptual Description of Complex Control System

We consider the problem of simulation of control system of complex multilevel dynamic systems having global control goal and consisting of a set of objects with their local control goals. As the system objective we consider the states, to which the objects should be transfered at some time interval. For this purpose, a control action corresponding to the current state is undertaken. This control action belongs to the set of control actions of the given object. The goals of objects at different levels of hierarchy are interconnected. Interaction between higher-level object and lower-level objects is such that the achievement of goals of one of the lower-level objects influences the achievement of goal of higher-level object. So, a set of control actions is correspondingly put to the goals of the object. This means that each of the objects or a class of objects is immersed onto the intersection of the states corresponding to the concrete set of the objects' goals, but the whole system is immersed into the state, called *hierarchical state* [12], which corresponds to the global control goal of the system. This approach is apppicable for any level of hierarchy. Hierarchical state S is a state, each component S^i of which corresponds to a set of parameters of certain level of hierarchy. Using the appropriate semantic interpretation, the hierarchical state shows how the current states of objects of different levels of hierarchy are related to each other.

The overall control problem is considered as construction of controlling scenarios realizing a time-event coordination of control actions that provide the achievement of control goal on the trajectories of system [11,12]. On each level an object is described in corresponding space of parameters and variables, some of which are polymorphic [12].

Based on the object-oriented model of control and problem domain, the system is presented as discrete-continuous system that reflects the process of state changes in the space of parameters. The parameters vary continuously on an observation time interval Δ. But the objects change their states discretely. This can formally be described as follows [12]. The state is defined as a set of trajectories of parameter changes at time interval Δ. The q-dimensional space of parameters is introduced, in which each object at time t is determined by the point $p(t) = \left(p^{(1)}(t), p^{(2)}(t), ..., p^{(q)}(t)\right)$. Each parameter is characterized by ν variables, $p = (\pi_1, \pi_2, ..., \pi_\nu)$. The i-th parameter at time t is represented by the point

$$p^{(i)}(t) = \left(\pi_1^{(i)}(t), \pi_2^{(i)}(t), ..., \pi_\nu^{(i)}(t)\right), \quad i = \overline{1, q}.$$

Then

$$p^{(i)}(t)_{t\in\Delta_j} = \left(\pi_1^{(i)}(t)_{t\in\Delta_j}, \pi_2^{(i)}(t)_{t\in\Delta_j}, ..., \pi_\nu^{(i)}(t)_{t\in\Delta_j}\right)$$

or simply

$$p^{(i)}(\Delta_j) = \left(\pi_1^{(i)}(\Delta_j), \pi_2^{(i)}(\Delta j), ..., \pi_\nu^{(i)}(\Delta_j)\right)$$

defines the trajectory that characterizes the dynamics of i-th parameter on time interval Δ_j. Then, the state S_j on time interval Δ_j is defined in space of parameters as

$$S_j = [p^{(1)}(\Delta_j), p^{(2)}(\Delta_j), ..., p^{(q)}(\Delta_j)].$$

Deformation of trajectories of parameters at different time intervals Δ_j and Δ_{j+1} formalizes the state transition, $S_j \rightarrow S_{j+1}$.

Discrete properties of the system is determined by the necessity to divide the state space into subspaces to reflect the observations that characterize the change of states upon transition from one subspace to another.

3   Dynamic Classification of Control Objects

The use of hierarchical state diagrams technique supposes the initial classification of control objects over the system state space and construction of the *canonical model of state dynamics* of a set of objects *Obj* under study.

The canonical model is constructed by using (1) databases containing the information about the parallel dynamics of values of parameters that characterize the set of objects at a chosen time interval, and (2) expert knowledge-base consisting of declarative and procedural knowledge reflecting the rules of classification. The classification rules are presented in the form of matrices with production rules, that are formulas of some language L, as elements. An element (i, j) of classification matrix, where i is a parameter and j is a class of objects,

$j \subseteq Obj$, contains a formula which determines the current process of i-th parameter dynamics at time interval. The matrices of this type give rules of one-level classification. However, the rules of *multilevel classification* are more important in hierarchical models. Multilevel classification is based on the gradual clarification of conditions that should be satisfied by objects from a class. In general case, the rules of multilevel classification are heuristic and reflect the knowledge and experience of problem domain experts. The subclass of multilevel classification rules which along with the grouping of objects reflect the semantics of state dynamics is of the most interest. To define multilevel classification, the notions of *state scale* and *classificator* are introduced.

Let $K = \{k_1, k_2, ..., k_q\}$ be a set of predicates, propositions, relating to the values of parameters of the objects' set Φ. The ordered set of predicates $K = \{K_1 < K_2 < ... < K_n\}$, $T_{K_i} \cap T_{K_j} = \emptyset$, $i \neq j$, where T_{K_l} is a truth domain of K_l, is called a one-level scale (further, scale) if each K_i defines a state S_i. We assume that predicates and the corresponding states have the same ordering, that is if $K_1 < K_2 < ... < K_n$ then $S_1 < S_2 < ... < S_n$. The scale determines the values of parameters and enables one to compare the states of the objects.

A scale $\{K_{i_1} < K_{i_2} < ... < K_{i_n}\}$ is the *hierarchical continuation* of the scale $\{K_1 < K_2 < ... < K_i < ... < K_n\}$ if the predicates $\{K_{i_1} < K_{i_2} < ... < K_{i_n}\}$ are the set of sub-predicates of K_i. A hierarchical system of scales is called to be a *classificator* of objects from Φ over the set of parameters at time interval Δ. The classificator is then used for formal description of state dynamics of the objects from Obj. The classification of objects by states to which they belong allows us to get a state space of the system, in particular, a state space $\widetilde{S}$ for the set of objects Obj; moreover, at each time interval Δ_j, $j = \overline{1, n}$, the system has its subspace of states $\widetilde{S}^{\Delta_j}$ and $\widetilde{S} = \cup_{j=1}^{n} \widetilde{S}^{\Delta_j}$.

A canonical model of state dynamics for a set of objects is represented by state transition diagram

$$D = \left\{ [0, \widetilde{T}], S, K, P^+, S_0, S^*, (\mu_i, i = \overline{1, n}, \mu_0, \mu^*) \right\}$$

where
$[0, \widetilde{T}]$ is a finite modeling interval; the interval is divided into n parts Δ_j such that $\cup_{j=1}^{n} \Delta_j = [0, \widetilde{T}]$ and $\cap_{j=1}^{n} \Delta_j = \emptyset$,
$S = \{S_j, j = \overline{1, n}\}$ - a set of states ordered by K; $S \subseteq \widetilde{S}$, $S_j \in \widetilde{S}^{\Delta_j}$,
S_0, S^* are initial and final states respectively,
P^+ is a set of arcs; each arc is assigned a time interval $\Delta \in [0, \widetilde{T}]$ of state transition, if $(S_1, S_2) \in P^+$ then $S_1 < S_2$,
$\mu_1, \mu_2, ..., \mu_n$ - distribution of objects over the vertices-states of the diagram at time moments $t_1, t_2, ..., t_n$ respectively, $t_j \in \Delta_j$; μ_0 - an initial distribution, μ^* - a final distribution.

The canonical model is used for formalization of qualitative properties of multilevel dynamical system and represents a hypothetical model (based on the expert knowledge) of state dynamics of a set of objects. The canonical model describes the qualitative character of state dynamics, using the distributions μ_i of objects Obj by determining the states to which they belong at time moments t_i.

The canonical model is used for comparative analysis with the actual state dynamics of a set of objects. For this purpose, the states of the objects under study are re-estimated at time moments $t_1, t_2, ..., t_i, ..., t_n$ in order to get the actual distributions of states of the objects over the states of canonical model, and then to compare this distribution with the required one. This helps represent the essence of system control problems and state dynamics of control objects. The description of actual process of state dynamics at arbitrary time interval Δ_j is based on the use of states of the canonical model as objects' classificator.

4   Hierarchical State Diagrams Model

The basic idea underlying the abstract representation of control process dynamics is to use hierarchical state diagrams. They provide qualitative description of dynamics of parameters and controllable state dynamics of objects with use of controlling scenarios.

The hierarchical state diagram is described by the model, each component of which has deterministic nature:

$$D^H = \left\{ [0, \tilde{T}], < D_A >, C, M, IM, H^c, CS \right\}.$$

The components are defined as follows.

$[0, \tilde{T}]$ is a finite modeling interval.
$D_A = < D, P^-, \eta(t), N(t) >$ is a state diagram model of actual state dynamics,

D is a set of canonical models for different groups of objects; P^- is a set of arcs, that describes a state backstep. Thus, the set P of D_A is defined as $P = P^+ \cup P^-$, $P^+ \cap P^- = \emptyset$, where P^+ is a set of arcs that describes state transitions. If $(S_i, S_j) \in P^+$ then $(S_i \prec S_j)$, and if $(S_i, S_j) \in P^-$ then $(S_j \prec S_i) \vee (S_j = S_i)$.

$\eta(t) = (\eta^1(t), \eta^2(t), ..., \eta^l(t))$, $l = |P|$, is a vector-function, called the counter of objects, each component $\eta^k(t)$ defines the number of objects that change their state from S_i to S_j, $k := (S_i, S_j)$, at some time interval $\Delta_{\alpha\beta} = [t_\alpha, t_\beta]$; the counters $\eta^k(t)$ are assigned to each arc $(S_i, S_j) \in P$. The counters assigned to the arcs P^+ characterize the intensity of the processes of objects' state transitions (positive processes); the counters assigned to the arcs P^- estimate the intensity of objects' state backsteps (negative processes).

$N(t) = (N^1(t), N^2(t), ..., N^n(t))$, $n = |S|$, is a vector-function, each component $N^i(t)$ defines the number of objects having a fixed state S_i.

The functions $\eta(t)$ and $N(t)$ enable one to obtain the information concerning the relation between processes of development and degradation, and to qualitatively estimate control actions and their efficiency.

C is a set of composition operations giving the rules of consistency of different state diagrams. These rules help consructing complex models of state dynamics that combine the requirements to the different sets of parameters and represent the conditions for coordination of state dynamics of the objects at different levels of hierarchy. Structural composition of state diagrams provides a synthesis of complex requirements set to the dynamical characteristics of controllable object. So, we say that for state diagrams the property of consistency holds if the attainability of certain states takes place in a given *time-event sequence*. As such operations the *sequential* and *parallel* compositions and operation of *generalization* are used. In case of generalization, for coordination of state diagrams at neighbor levels of hierarchy the Cartesian product of states of diagrams of lower level of hierarchy can be defined. In this case, one should specify the ordering relation on the subsets of Cartesian product of states of diagrams of lower level of hierarchy. The composition of state diagrams allows one to formally represent different combinations of complex criteria sets in order to perform objects classification and to solve control problems. Using the consistency rules and operations with state diagrams one can model diverse schemes of inter-level relations and influence of states of lower level diagrams on the processes of higher levels of hierarchy. As a result, a certain value is produced at the highest level output, which is considered as a response of hierarchical system on the input control symbols.

M is a set of output results m of monitoring and/or diagnosing of control object, $m = \{m^1, m^2, ..., m^r\}$.
$IM = \{DB, KB, M, PR\}$ is the information-mathematical model describing the behavior and dynamics of objects at time intervals between the events:

DB - databases.

KB - knowledge-bases.

M - analytical models of object' dynamics at time intervals.

PR - production rules constructed by experts and based on DB and KB; they give rules for re-estimation of current situation in dynamics of parameters and possible formation of new states.

$H^c = \{S, P, S_0, S^*, X, V, F\}$ is a model of controllable state dynamics that illustrates the key dynamic characteristics of system depending on whether or not the control actions corresponding to the current states are performed. The model is represented in the form of state transition diagram with the following components:

S is a set of states.

S_0, S^* are initial and final states, respectively.

X is a set of input control symbols.

P is a set of arcs, $P = P^+ \cup P^-$, $P^+ \cap P^- = \emptyset$. P^+ is a subset of arcs that define state transitions initiated by input control symbols, P^- is a subset of arcs that define state backsteps when no control symbols are entered the input.

$X \Leftrightarrow P^+$ is a one-to one correspondence that defines for each input control symbol the state transition initiated by this symbol.

$V = (V_x, V_m, V_{im})$ are the operators of synthesis of new initial conditions, new states and object' dynamics depending on the control actions, the information about the results of monitoring or diagnosing, and the results of changes in IM, respectively.

F is a connection rule between state transitions at neighbor levels of hierarchy. The rule is defined by the mechanism of after-effect by splitting P^+ and X into two subsets (P_z^+, P_u^+) and (X_z, X_u), respectively. The arcs of P_z^+ are called *isolated*, and the arcs of P_u^+ are called *coupled*. In accordance to this partition, control symbols of X_z are called *individual*, and control symbols of X_u are called *general*. The coupled arcs are defined by introducing the parent-arcs as a Cartesian product of child-arcs of state transition diagrams at neighbor levels of hierarchy. The isolated arcs P_z^+ describe the state transitions initiated by individual input symbols X_z; this kind of symbols do not influence the state transitions of other diagrams. The coupled arcs P_u^+ describe the state transitions initiated by general input symbols X_u; this kind of symbols initiate the state transition on the parent-arc, which means, as a consequence, the state transitions on the corresponding child-arcs. And conversely, state transitions on all/several child-arcs initiate a state transition on the parent-arc of the diagram of higher level of hierarchy.

$CS = \{\Omega, I, H, T, A\}$ is a model of control scenario of state dynamics of objects:

Ω is a set of state transition diagrams; they describe the state dynamics for each object.

I is a hierarchical structure.

$H : I \to \Omega$ is a functional that assigns a hierarchical number to each diagram of Ω.

T is a time diagram for symbols X; it determines the sequential-parallel process of input control symbols entering.

A is a scheme of after-effect of state transitions.

To give the time diagram T, different ways can be used, including the estimation rules of each current state of the system.

The trajectory of attainable states represents general and local goals solved by scenario on arbitrary time interval. The study of basic properties of control scenario is based on the analysis of the trajectory of attainable states and its comparison with the expected or predicted effect.

5 Automated Computer Simulation System

The developed technique is realized in automated computer system of simulation and analysis of dynamical processes, scenario modeling and control of complex hierarchical systems. The automated computer system is based on the concept of comprehensive use of information systems and technologies. The system has a modular structure that provides sufficient convenience and facility of editing of the separate modules, not influencing the functioning of the others, and adding of new functional possibilities.

The information on the problem domain and properties of the object under investigation is contained in specialized databases; knowledge about parameters and processes is contained in knowledge bases; information about the current values (dynamics) of parameters and on the character of state dynamics of object is contained in monitoring databases. The investigations on the models of simulation technique are provided by the combination of the methods of (1) dynamic expert systems, (2) production expert systems, (3) database processing techniques, (4) monitoring/diagnostic data analysis, (5) scenario control and modeling.

Knowledge of values of various parameters and their change is one of the most important elements providing adequate representation and modeling of state dynamics of complex systems. Monitoring allows one to carry out observing for the current values of parameters and for the actual information on the character of system dynamics. This information is further used to evaluate conditions/situations around a system and to predict probable events in the system and consequences following from them, which can be caused by changes in values of parameters.

The architecture of computer system consists of several subsystems, logically related to each other.

I. The subsystem of *direct planning* includes:
   (a) *User interface* provides interaction of user with the system. The software of user interface serves as a tool that realizes all functions of computer system associated with information exchange with the user.
   (b) *Library of parameters* contains blocks of parameters for supporting the continuous process of observing for a number of parameters; the library is extendable and editable.
   (c) *Knowledge-base* is a computer realization of formalized expert knowledge about problem and control domains.
   (d) *Builder of canonical model of state dynamics* is a specialized module of entering of state diagrams as input information. The state diagrams tool provides clear and precise formalization of states, inherent for one process and not typical for others. It can be used for representation of regularities and typical models of state dynamics.
   (e) *Monitoring databases*. Direct planning in control systems assumes a high level of informatization and operative connection with monitoring database.
   (f) *Interpreter of monitoring database*. The interpreter of monitoring database and the model of controllable state dynamics are the basic components in automated computer system. The interpreter of monitoring database functions according to the composition of canonical models of state dynamics specified by the user and generates the description of actual multilevel dynamics of the object.
   (g) *Model of controllable state dynamics* is constructed as expert subsystem for qualitative estimation of control scenarios defined by the user.

The subsystem of direct planning, in a generalized form, realizes the two-stage process of simulation and control. The first stage is the stage of *retrospective analysis* consisting in the construction of *predictive model of inertial state dynamics*, that is the model of dynamics of object when no control actions are undertaken. The retrospective analysis provides the means of events prediction. When predicting events, the parameters of system are continuously measuring. If there was some event in a system and for some time before the event a parameter has sharply changed, or there was a gradual change of values of parameter up to some critical, then such anomaly is related with this event. The dependences of such a kind confirmed repeatedly, becoming stable, are used for estimation and prediction of possible future events in the system. Actually, knowledge and experience obtained in the past and expert knowledge are used. The stage of retrospective analysis provides user with the tools of selection of objects, study of chosen parameters, and construction of state diagram, which interprets the monitoring data. This implements the diagnostic analysis of objects' state dynamics. By analysis, experimentation, and selection of different sets of parameters, the diagrams that most expressively depict "negative" (positive) trends are found. These diagrams formally represent the current control problems and answer the question "what will be if no control actions are performed". The results of retrospective analysis help put forward the goals and control problems and to form possible alternatives of controllable state dynamics for the perspective period.

The second stage consists in construction of the *model of controllable state dynamics* which includes *statement of control problem, alternative control scenarios, expert estimation and selection of control scenarios* of object's state dynamics. The stage includes the construction of the model of controllable state dynamics, analysis of controllable processes of objects' state dynamics, and obtaining the answer to the question "what control actions should be undertaken to achieve the required goals". At this stage an initial state and expected final state are described, and the space of intermediate states are constructed. Then, the conceptual model of control scenario in the form of *States Generator* is defined.

II. The subsystem of *multilevel dynamics simulation* includes:

(a) *The system of canonical models of state dynamics of hierarchical object* provides the necessary tools for construction of canonical state dynamics models of objects under study.

(b) *Scheme of coordination of canonical models* is based on the rules of composition of state diagrams.

(c) *Description of actual state dynamics of hierarchical object* uses the monitoring databases that store the information on the actual dynamics of parameters.

(d) *Monitoring databases* contain the current information about the dynamics of parameters, which is the result of monitoring or diagnosing of the object.

(e) *Partial/incomplete canonical model of state dynamics.* The model of controllable state dynamics is used for checking a hypothesis about the efficiency of scenario being estimated. The scenario estimation criterion is given in the form of "partial" or "incomplete" state dynamics diagram determining support states that should be reached with the specified restrictions on time and resources.

The special case of "partial/incomplete" state dynamics diagram is the pair of states: initial state and desirable final state. In this case, the expert subsystem should: *either* confirm a hypothesis that the model of state dynamics, controlled on the basis of scenario, meets the given criteria or requirements, and supplement the input diagram with the specifying intermediate states, *or* refute the hypothesis and generate computer prediction in the form of alternative state dynamics diagram.

(f) *Scenarios of state dynamics/Conceptual model of scenarios - states generator*. The Scenario of State Dynamics serves as an inference system. It is considered as generator of consecutive states of object under investigation. The rules of States Generator are represented in the format of tree-like decomposition of global goal on the sub-goals; to each terminal node an elementary rule is assigned. The rules are represented as follows:

$$(S_i, S_k, S_l, u_{ik}, R_{ik}, t_{ij}),$$

where u_{ik} is a control action needed to take an object from current state S_i to the subsequent state S_k, using the amount of resources R_{ik} and time t_{ik} and not admitting the backstep to the previous state S_l.

Using the toolkit of dynamic expert systems, the information environment can be adapted to the current range of problems with minimal costs.

III. The subsystem of *simulation process control* includes:

(a) *Executive module* consists of *database of objects' control models, database of state dynamics simulation models, database of analyzable control actions, simulation system of controlling scenarios, database of complex control actions*. Simulation system of controlling scenarios that uses the database of objects' control models, the database of state dynamics simulation models, and the database of analyzable control actions provides a tool for formal representation of goals, control problems and state dynamics, time and resource characteristics.

The simulation system of controlling scenarios enables user to analyze global efficiency of control actions directed on the achievement of goals at different levels of hierarchy. The system allows user to not just compare separate control actions but also to formally synthesize complex control actions (control strategies) for hierarchical system as a whole, and then their further comparison. The functional submodules of Executive Module reproduce the stages of control actions selection, including the stage of synthesis and initiation of problem domain and control models.

(b) *Control module* realizes the functions of expert estimation and comparison of control action sets, knowledge-base support, and decision making.

The simulation process control subsystem provides the tools of synthesis, simulation, and analysis of control scenarios.

The computer simulation system provides:

- Identification and registration of the information about the occurred events and the current situation
- Information storage and maintaining
- Information management by gathering, aggregations, classifications, processing, and delivery of necessary and requested information.

The rules of state transformations and the tools of construction of control scenarios

- Allow one to easily realize iterative process of creation and modification of control scenarios
- Admit the efficient realization by means of executive procedure
- Possess the sufficient expressiveness of specification of control processes.

The State Generator enables user:

- to study effects of integrated and multi-aspect control regarding different objects of complex system
- to divide the control process into stages
- to perform decomposition schemes of prediction, in which each subsequent model is an integrated or detailed elaboration of the previous
- to construct and analyze the interconnected aggregated and detailed models of state dynamics parameters.

6   Conclusions

The proposed simulation technique can be used for design of applied weakly-formalized control systems for simulation, analysis, control, and prediction of state dynamics in complex dynamical hierarchical systems. The technique presented can also be used as a technology for construction of computer systems for simulation analysis of development strategies and control scenarios of complex objects.

The proposed simulation models and technique are universal and at the same time problem-oriented in relation to the rationality, consistency and coordination of control actions. It can be equally used for diverse variety of systems such as technical systems, organizational systems, socio-economic systems, systems of strategic planning, and management information systems and decision support systems.

Acknowledgement. This work was supported by the Security Engineering Research Center, granted by the Korea Ministry of Knowledge Economy.

References

1. Bar-Yam, Y.: Dynamics of Complex Systems. Addison-Wesley, Reading (1997)
2. Caines, P., Wei, Y.J.: The hierarchical lattices of a finite state machine. Systems and Control Letters 25, 257–263 (1995)

3. Forrester, J.: Industrial Dynamics. Productivity Press, Portland (1961)
4. Pappas, G., Lafferriere, G., Sastry, S.: Hierarchically consistent control systems. IEEE Trans. Automatic Control 45, 1144–1160 (2000)
5. Harel, D.: Statecharts: A visual formalism for complex systems. Science of Computer Programming 45, 231–274 (1987)
6. Heimdahl, M., Leveson, N.: Completeness and consistency in hierarchical state-based requirements. IEEE Trans. Software Engineering 22, 363–376 (1996)
7. Mesarovic, M.D.: Theory of hierarchical multilevel systems. In: Mathematics in Science and Engineering. Academic Press, New York (1970)
8. Wong, K., Wonham, W.: Hierarchical control of discrete-event systems. Discrete Event Dynamical Systems 6, 241–273 (1995)
9. Wong, K., Wonham, W.: Hierarchical control of timed discrete-event systems. Discrete Event Dynamical Systems 6, 275–306 (1995)
10. Zhong, H., Wonham, W.: On the consistency of hierarchical supervision in discrete-event systems. IEEE Trans. Automatic Control 35, 1125–1134 (1990)
11. Bagdasaryan, A.: System theoretic viewpoint on modeling of complex systems: design, synthesis, simulation, and control. In: Zaharim, A., Mastorakis, N., Gonos, I. (eds.) Recent Advances in Computational Intelligence, Man-Machine Systems and Cybernetics, pp. 248–253. WSEAS Press, Stevens Point (2008)
12. Bagdasaryan, A.: Mathematical and computer tool of discrete dynamic modeling and analysis of complex systems in control loop. International Journal of Mathematical Models and Methods in Applied Sciences 2(1), 82–95 (2008)

The Realization of Greenhouse Monitoring and Auto Control System Using Wireless Sensor Network for Fungus Propagation Prevention in Leaf of Crop

DaeHeon Park, SungEun Cho, and JangWoo Park

Sunchon National University, department of information and communication Engineering
{dhpark,chose,jwpark}@sunchon.ac.kr

Abstract. The precision agriculture manages the environmental condition in streamlined way with the information from the plant, soil and greenhouses. And the greenhouse management system using wireless sensor network (WSN) is being developed to improve the productivity of the agricultural products. In this paper, we implemented a greenhouse monitoring and automatic control system in order to prevent proliferation of mold before the leaves of plants have become damp. To achieve this goal, we installed sensor node with the sensor for ambient temperature, relative humidity, temperature of leaves, humidity of leaves, and illumination from the inside of the greenhouse and rain sensor on the outside. We applied the formula of Barenbrug dew point in this experiment to calculate the dew point with the gathered data from the sensor node. To prevent dew on the surface of leaves, an algorithm of the environment server which manages the environment control device with the calculated dew point was used. We have installed sensor node to the miniaturized greenhouse model and verified our greenhouse monitoring program by observing the greenhouse condition.

Keywords: Greenhouse, Auto-Control, Leaf sensor, Dew point, Sensor network.

1 Introduction

These days in many countries and various fields are trying to apply sensor networks to the parts of agricultural production and management [1].

Interest is on the precision agriculture, managed apart as growth or soil condition in the greenhouse. For improvement in agricultural productivity, sensor network should be applied to control skill of greenhouse system, database artificial intelligence and data communication system [2].

When the leaves of crop temperature decreased to below sew point, the mold is apt to grow as moisture in the air becomes water drops on the surface of crop leaves [3][4].

The proposing system in this paper is the system preventing mold propagation by using dew point for prevention forming dew on the surface of leave of crop.

To monitor greenhouse environment we installed sensor node with the sensor of air temperature, air humidity the temperature of leaves, humidity of leaves and illumination in the inside of the greenhouse and rain sensor in the outside.

D. Ślęzak et al. (Eds.): CA 2009, CCIS 65, pp. 28–34, 2009.

In greenhouse, we applied formula of Barenbrug dew point to this experiment for creating the environment with dew and calculated the dew point that used the sending data air temperature, air humidity, temperature of leaves from sensor node.

In order not to go down below dew point, the system controlling automatically greenhouse environment was designed.

The construct of this paper is as follows. In section 2, explain the fungus propagation prevention system in the greenhouse. In section 3, realize proposed system and then check up the results of experiment. Finally in section 4, describe the results.

2 Fungus Propagation Prevention System in the Greenhouse

We calculated the dew point after measuring the air temperature, the air humidity and the temperature of leaves.

The temperature of leaves, to prevent forming dew on the surface of leaves of crop. The fungus propagation prevention system is that to maintain proper humidity when we manage the cultivation of the crops.

2.1 A Component of Fungus Propagation Prevention System

Fig. 1 is the configuration of the fungus propagation prevention system. It consists of sensor group, control group, management group and monitoring group. Sensor nodes installed in the greenhouse (air temperature, humidity sensor, temperature of leaves, humidity of leaves and rain sensors) collect environmental data. The collected data is sent to the base node the data, base node send to the environment server sent from each sensor. Using data collected from the sensor, the environment server send a signal to the relay node control signals according to the preferences the dew calculates and the administrator wants to. Relay node operates each devices according to settings transmitted from the environment server. Greenhouse data transmitted from the

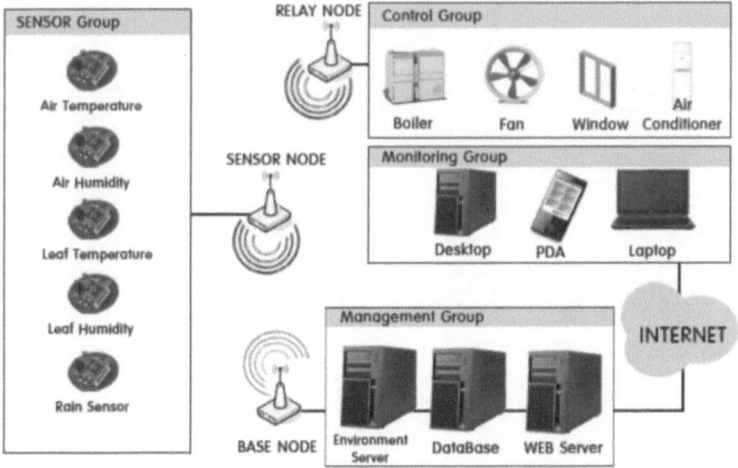

Fig. 1. Overall System Diagram

environment server is stored in the database server, through a Web server, the Internet and the administrator's platform (laptop, PDA, desktop) can be monitored the information of the greenhouse condition.

2.2 Leaves Temperature Sensor

To measure the exact temperature of leaves, we designed the leaves temperature sensor using RTD (Resistance Temperature Detector) temperature sensor. Existing Leaves temperature sensor used the non-contact infrared sensor, but it was difficult to measure the exact temperature of leaves because amount of current is very little. Leaves temperature sensors developed in Phytalk company is contact style temperature sensor, but it is very expensive and hard to get in domestic. So this system measures the temperature of the leaves using RTD temperature sensor in contact with directly to the surface of the leaves of crops according to the temperature change of leaves of crops converting resistance value of the RTD sensor into temperature value. RTD sensor has a error value of ± 0.3 degree.

Fig. 1 shows the temperature characteristic graph of the PT1000 sensor. Temperature characteristics graph was formed using the following formula matlab[5].

$$- 50℃ \sim 0℃ : R(t) = R0(1 + At + Bt^2 + C(t - 100) \times t^3)$$
$$0℃ \sim 50℃ : R(t) = R0(1 + At + Bt^2) \tag{1}$$

A : 3.9083x10-3x℃$^{-1}$, B : -5.775x10-7℃$^{-2}$, C : -4.183x10^{-12}℃$^{-4}$

R0 : resistance value in ohm at 0℃

t : temperature

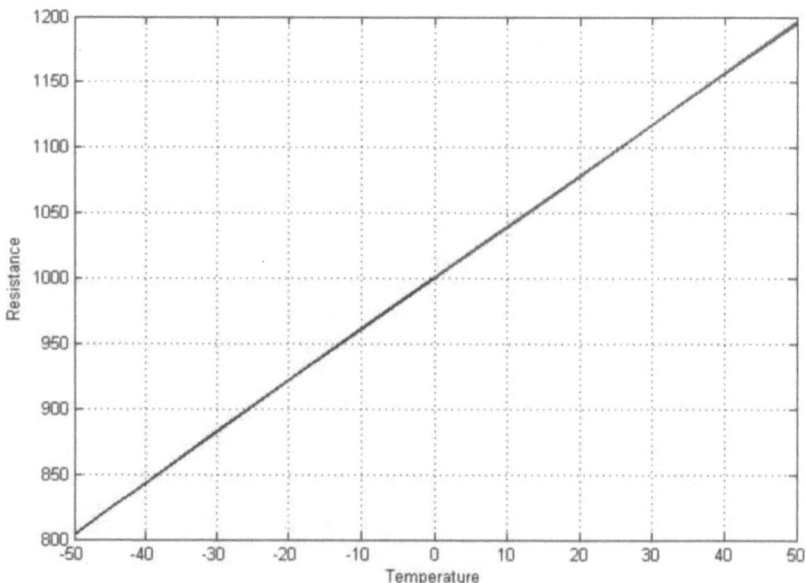

Fig. 2. Characteristic temperature curve

2.3 Dew Point

In order to calculate the dew point, environment servers are transmitted from Sensor node the value of the air temperature, humidity, temperature of leaves and humidity of leaves.

It substitute the Barenbrug dew point formula for Transmitted data. This equation has the error value of [6].

It is valid for:

$$0\,°C < T < 60\,°C$$
$$0.01 < RH < 1.0 \tag{2}$$
$$0\,°C < Td < 50\,°C$$

Where

T - Temperature in degrees Celsius
RH - Relative humidity
Td - Dew point temperature

The equation are:

$$T_d = \frac{b \cdot \alpha(T, RH)}{a - \alpha(T, RH)} \tag{3}$$

Where

$$\alpha(R, TH) = \frac{a \cdot T}{b + T} + \ln(RH/100) \tag{4}$$

a = 17.27, b = 237.7 °C (a, b is constant)

2.4 The Fungus Propagation Prevention System Flowchart

Fig. 3 displays the fungus propagation prevention system Flowchart. If the sensor data is collected from the base node, the fungus propagation prevention systems calculate the dew temperature based on this data, detect environment formed dew as dew temperature is higher than temperatures measured in the temperature of leaves sensor.

Sensor data is collected at intervals of five seconds by the sensor nodes, using it, if the dew temperature increases more than temperature of leaves, the sensor data is collected, and then these as a timer the value after increasing the counter. (at. Counter 1 per 5 seconds: Counter 360 = 30 minutes, applied algorithm is at interval 30 minutes.)

If the dew temperature is higher than the temperature of leaves and any timer was not started, the first timer is started as the boiler is run. And then it is stored in database, after that wait for the next data. User set specified time the timer expires and check periodically whether the timer has expired or not. After running the first timer, if the dew temperature is higher than the temperature of leaves, the first timer is expired, on this occasion; the second timer is started, at the same time the ventilation fan is operated.

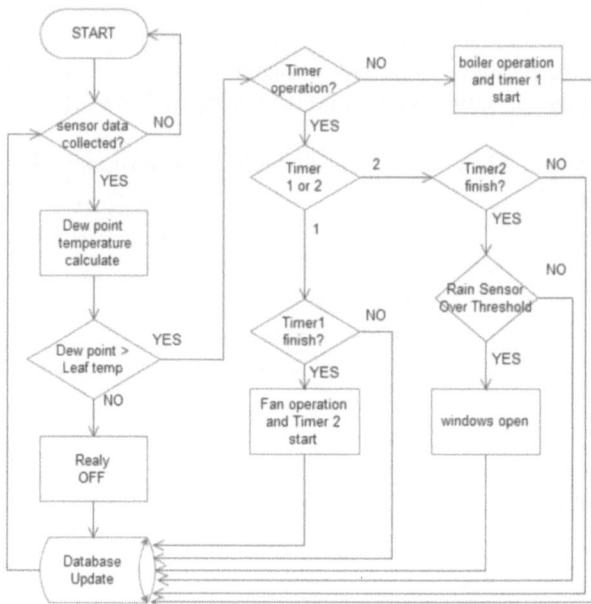

Fig. 3. Overall system Flowchart

As the second timer is expired, Rain Sensor concludes whether to open the window after detecting whether it rains or not outside. As the dew temperature is lower than temperature of leaves, all the Relay operated is the OFF state.

3   System Implementation

To verified the performance of the greenhouse monitoring system after making a model of the greenhouse as follows figure 4.

The information collected from environment sensor installed in the model of greenhouse is stored in environment database, and it can be provided GUI screen to the users. The various information about various disease of facilities crops are required. This paper provides the service to make environment which the dew isn't formed using environment data (the air temperature, humidity, the humidity of leaves, the temperature of leaves).

There are the graph of the dew point temperature which is compared the dew point calculated with the temperature of leaves in real-time, and the graph of environment temperature which is shown changes in temperature of environment of greenhouse.

Fig.5 is GUI screen displaying the information data collected from each sensor nodes. With the GUI, a user can change the setting value of greenhouse and directly can control the control device of greenhouse.

Fig. 4. Greenhouse model & measurement temperature of leaves sensor

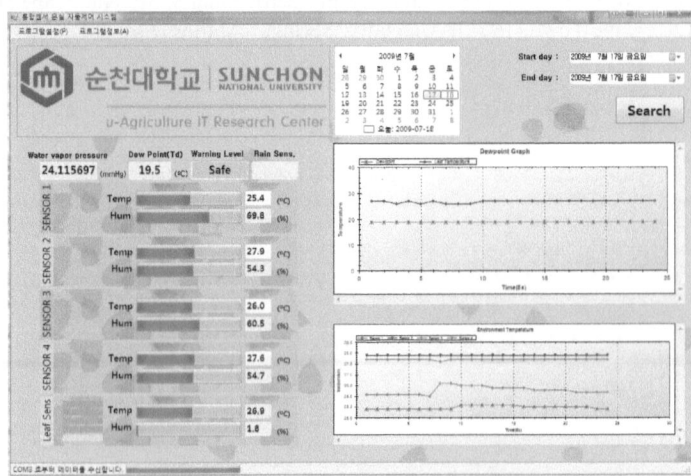

Fig. 5. Greenhouse Monitoring System

4   Conclusion

In this paper, this sensor provides information to the user after grasping the condition of the greenhouse environment and crop using it to every time.

Server optimizes using the environment server to maintain certain humidity and the optimal value of the humidity to the growth of crops in the greenhouse.

Also, it embodied the fungus propagation prevention system calculating the dew point in order to prevent forming dew on the surface of leaves of plants.

To verify the performance of this system, by create a virtual model of the greenhouse, equipped with environmental sensor, monitored the condition of environment through monitoring the greenhouse.

Acknowledgements. This research was supported by the MKE(Ministry of Knowledge Economy), Korea, under the ITRC(Information Technology Research Center) support program supervised by the IITA(Institute for Information Technology Advancement)" (IITA-2009-(C1090-0902-0047)) and this work was supported in part by MKE & IITA(09-Infr, industrial original technology development project).

References

1. http://www.nia.or.kr/open_content/board/
   boardView.jsp?id=28881&page=3&tn=KY_0000219
2. Kim, Y.-S.: Expert Development for Automatic Control of Greenhouse Environment. J. Kor. Flower Res. Soc. 12(4), 341–345 (2004)
3. Choung, B.-M.: u-Farm Foreign application case book. Nation Information Society Agency (2006)
4. Kang, G.C., Yon, Ryou, K.S., Kim, Y.S., Paek, Y.J., Kang, Y.K.: Effects of Humidity Environmental Control in Greenhouse Using Refrigeratory-Based Dehumidifier. The Korean Society for Bio-Environment Control Soc., 149–153 (2006)
5. http://www.ogamtech.com/s/SUM_TCTB01-101,1001_E.pdf
6. Barenbrug, A.W.T.: Psychrometry and Psychrometric Charts, 3rd edn. Cape and Transvaal Printers Ltd., Cape Town (1974); Appendix: Springer-Author Discount

Selection and Optimization of Alternative Modular Products Using Evolutionary Computing

Orlando Durán[1], Nibaldo Rodriguez[1], and Luis Perez[2]

[1] Pontificia Universidad Católica de Valparaiso, Valparaíso, Chile
{orlando.duran,nibaldo.rodriguez}@pucv.cl
[2] Universidad Técnica Federico Santa María, Valparaíso, Chile
luis.perez@usm.cl

Abstract. In most configurations of modular structures, products are assumed to have a unique modular structure. However, it is well known that alternatives for constructing modular structures may exist in any level of abstraction. Explicit considerations of alternative structures invoke changes in the number of module instances so that lower costs, more independency of structures and higher efficiency can be achieved. Relatively few research papers deal with the optimization of modular structures problem with alternative assembly combinations aiming at minimization of module investments was found in the literature. A genetic algorithm (GA) was applied to solve the optimization problem of selecting and combining the alternative of modular structures to create a set of modular structures minimizing the cost involved in its implementation. Test results are presented and the performance of the proposed GA and compared to Simulated Annealing technique and compared to solutions obtained from total enumeration tests. Finally an evaluation of the performance of the proposed G.A. in a set of large problems is presented and discussed.

Keywords: Modularization, Genetic Algorithms, Modular structures.

1 Introduction

Market competitiveness forces many industries to evolve toward mass customization (MC), which aims at satisfying individual customer needs while keeping mass production efficiency [1]. Modularity is a well-established strategy for attaining MC. Baldwin and Clark [2] defined modularity as a concept that is applied to manage complex systems by breaking them down into parameters and tasks that are interdependent within and independent across the modules. Modularity is one of the primary means of achieving flexibility, economies of scale, product variety and easier product maintenance and disposal. One of the major trade-offs for transformation into modular systems then becomes to select the combination of modules to assemble. The characteristics of modularity comprises the use of a finite set of components to meet the infinite changes of the environment; establish the module by reviewing the similarities

D. Ślęzak et al. (Eds.): CA 2009, CCIS 65, pp. 35–46, 2009.

among the components; keep as much independence of the resulting structures as possible and use different modules for different varieties of assemblies.

The replacement of dedicated structures by modular ones seems to be a trend in the manufacturing field, especially in meeting the desire for greater flexibility. Products built around modular architectures can be more easily varied without adding too much complexity to the manufacturing system, and moreover, a modular architecture makes the standardization of components more possible. At the same time this approach allows the reuse of tools, equipment and expertise, and avoids costly changeovers for personalized products. According to Pandremenos [3] there are three general fields where modularity could be implemented: modularity in design, modularity in use and modularity in production. Despite these clear benefits, a formal theoretical approach to modularity is still lacking and designers are often skeptical regarding the advantages of modularity. This is largely due to the inferior performance obtained by modular designs compared to their custom built optimal alternatives. Determination of modular configuration is described in [4] as, "Given a set of candidate modules, produce a design that is composed of a subset of the candidate modules and which satisfies both a set of functional requirements and a set of constraints". From this definition, it can be seen that here we assume that modular architectures of a particular product of structure is ready, and modular components and their interactions are predefined and available. There are many domains where a wide variety of modular designs are available. Many engineering problems can be generalized under the umbrella of modularity [5]. Depending on the specific domain at hand, many researchers have reported automated systems and methodologies for define one or more modular configurations for a given application, i.e. Asan [6] et al. developed an integrated method for designing modular products. To test and validate the methodology it was applied to a domestic gas detector product family. Hornby [7] developed an automatic design system that produces complex robots by exploiting the principles of regularity, modularity, hierarchy, and reuse. Babu [8] developed an automated fixture configuration design system to select automatically modular fixture components for prismatic parts and place them in position with satisfactory assembly relationships. Finally, Retik and Warszawski [9] proposed a knowledge-based system for the detailed design of prefabricated building. Benefits of a standardized modular design approach overtake customized approaches. Modularity attributes are influenced by the technologies used in the modular components, the work agents that perform the assembly, and their supporting infrastructure ([10], [11]). There is more than a unique form to build a given solution or a structure using a set of modular components from a given and a finite set. The definition of a modular alternative requires comparative estimates of time, performance and cost among alternative of modular configurations.

1.1 Problem Statement

Consider a simple example, a lego-based figure, where a finite number of alternatives of combinations, using modular components, are possible for a given figure. Modular systems are comprised of many modular elements and modules that can be stacked in a flexible way. See figure 1, where two alternatives of combinations to construct the same figure are shown. In figure 1(a) three components were used and in figure 1(b) two components were used. Each modular component has its known cost. If number of figures is needed to be assembled at the same moment, and several modules are common to some of these figures, and, if there is a limited number of each one of the modular components, the challenge here is selecting the optimum combination of modular designs for each one of the lego-like figures at a minimum cost.

(a)   (b)

Fig. 1. One Two alternative configurations for constructing the same lego-like figure

This paper proposes a method which helps companies change their dedicate systems into modular systems. The optimization is achieved through appropriately selecting the subsets of module instances from given sets. In general, each module may have more than one instance and any each assembled product may have more than assembled combination of modules. The different alternative assemblies may provide the same capabilities and even functionalities that the required ones. Furthermore, current work may accommodate simultaneity constraints where the selection of a particular instance of a module necessitates the use of a particular instance of another module. The method is applied to a general problem and is found to be efficient in determining optimum subsets of each module from a given set. It is anticipated that the proposed method can be used as a systematic tool in selection of modules instances in designing and assembling modular products or transforming dedicated structures into modular ones. In the case of a custom made product the manufacturing system should be set up to produce a different product variant for each parameter set, representing a different specification, besides the large number of components that are needed to comprise all the combinations. In a modular-oriented environment, the number of required modular instances may be significantly reduced since modularity provides the desired variety of the product through different combinations of modules. Two key issues in modular product design are to determine the optimum product variety and the number of module instances required to support this variety. For each parameter set, the best possible product variant is selected as a result of the best combination of module instances. The problem is to select the subsets of modules instances that minimize a combined cost objective function.

2 Genetic Algorithms and Simulated Annealing

The basic concepts of Genetic Algorithms (GA) were developed by Holland [12]. Holland showed that a computer simulation of this process could be employed for solving optimization problems. Every solution to the addressed problem is called a chromosome. A chromosome is a string of binary bits or numbers. The pool of solutions is called the population. New candidates are generated gradually from a set of renewed populations by applying artificial genetic operators selected from strategies based on the survival of the fittest principle, after repeatedly using operators of crossover and mutation. Genetic Algorithms have been used to solve a variety of optimization problems with some success. Combinatorial problems constitute a class of problems particularly difficult to solve, sequencing problems included. GAs need only a fitness or objective function value. No derivatives or gradients are necessary. Finally, GAs use probabilistic transition rules to find new design points for exploration rather than using deterministic rules based on gradient information to find these new points. A random initial population initiates the evolutionary process. Each chromosome is assigned a positive value called fitness proportional to the quality of the solution. Based on their fitness value, subsets of the chromosomes of the current population are selected for reproduction. The reproduction is accomplished by applying a set of genetic operators, called crossover and mutation, to the chromosomes.

The basic algorithm of the GA is given as follows:

1. Generate random population of chromosomes.
2. Evaluate the fitness of each chromosome in the population.
3. Test if the end condition is satisfied, stop and return the best solution.
4. Create a new population by repeating following steps until the new population is complete:
   - Reproduction: Select two parent chromosomes from the population according to their fitness.
   - Crossover: With a crossover probability, crossover the parents to form a new offspring (children).
   - Mutation: With a mutation probability, mutate new offspring.
5. Replace: Use new generated population for a further run of algorithm.
6. Go to step 2.

Simulated Annealing (SA) was proposed by Metropolis [13] and it has been the object of intensive analysis and implemented to solve a series of optimization problems. The main steps of the SA algorithm can be summarized as follows: It starts with an initial feasible solution, repeatedly generate neighbor solutions. A generated solution is accepted if it has a better value of the objective function. If it is worse, the solution may be accepted depending on certain probability distribution. The acceptance probability depends on the cooling temperature T. At the beginning, T is set to a value which allows the acceptance of a large portion of the generated solutions. Then T is modified to reduce the acceptance probability. This fact assures that the algorithm does not converge to local optimum at early steps. Several are attempted at each vale of T and the algorithm is stopped as soon as a stopping criterion is reached.

3   Mathematical Formulation

In this paper, we attempt to solve a generalized selection and optimization problem in modular construction of a set of figures. Consider a situation having 'F' figures and 'M' modules. Each figure is characterized by its structure, and each structure is comprised for a given combination of different modules. Parts may have S alternative structures and all parts are not having equal number of alternative structures. Under a certain modular configuration for a figure there is associated with a given cost, in correspondence to the total number and types of modules to construct the referred structure or figure. Thus the problem is to find optimal parts configurations with their modules combinations to minimize total costs. The mathematical model for cell formation problem, in the presence of multiple modular configurations is presented below. As it was mentioned, the problem consists in determining the minimal cost required for assembly all the figures simultaneously, selecting the appropriate combination of alternative assembly for each one of the figures. Here, all the figures are to be assembled simultaneously which it means that the need of some modules could be incremented according to the number of figures that use a specific assembly option that considers the module at hand. Let us introduce the elements of this optimization model. The optimization model is stated as follows. Let:

- M be the number of modules,
- F the number of figures,
- S the number of alternative assemblies or set ups,
- i the index of figures (i = 1,....,F),
- j the index of modules (j =1,.....,M),
- k the index of alternative assemblies (k=1,.....,S),
- A = [a_{ij}] the F x M binary incidence matrix,

Given an incidence matrix F= [f_{ijk}], where:

$$f_{ijk} = n,$$

Where n represents the number of modules of the type j which are used by the figure i in the alternative of assembly k. Each module j has a unitary cost c_j. We selected as the objective function to be minimized the cost of a given set of set up of modules to assembly a set of figures to be constructed simultaneously.

$$Min : Z = \sum_i^f \sum_j^m \sum_k^s F_{ijk} C_j A_{ik}$$

$$A_{ik} = \begin{cases} 1 & \text{if figure i will be assembled using the assembly alternative k} \\ 0 & \text{otherwise.} \end{cases}$$

Subject to

$$\sum_{k=1}^C A_{ik} = 1 \qquad \forall i,$$

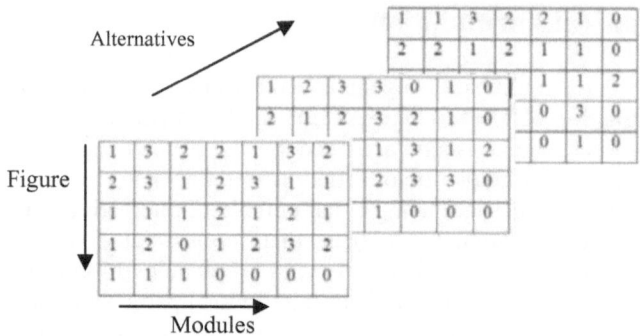

Fig. 2. Generic structure of the selection and optimization problem

The first and most important step in preparing an optimization problem for a GA solution is that of defining a particular coding of the design variables and their arrangement into a string of numerical values to be used as the chromosome by the GA. In order to generate the chromosomes, the length of the chromosome is calculated first. Then random numbers in the range of {0, S} are generated to form the chromosome. In this paper, a direct coding scheme is used, that is the allele of each gene represents the alternative of module subset to which each part is constructed. The chromosome representation is shown below where Mj denotes the alternative to which figure j has been assigned.

M1   M2   M3   . . .   MM

For example, consider the following 3 chromosomes where figures can be comprised by 10 modular components in 4 alternative assemblies maximum.

1	2	4	2	3	3	1	2	3	2
2	1	2	1	1	2	3	3	1	1
2	2	1	2	1	2	3	3	4	1

The first chromosome indicates that figure 1 is constructed using its modular alternative number 1, as well as figure number 7. As such, figures 2,4,8 and 10 are considered as using its alternative number 2. Figures 5,6 and 9 are constructed according to alternative structure number 3. Figure number 3 is constructed according the alternative structure number 4. Note that the chromosomes are fixed in length based on the size of the problem that is number of figures or structures considered as to be assembled simultaneously. There are many different types of reproduction operators, which are proportional selection, tournament selection, ranking selection, etc. By selection, crossover and mutation it is possible to obtain new population from the current one. In this study, reproduction is accomplished by copying the best individuals from one generation to the next. This approach is often called an elitist strategy. For the number of elitist individuals we used a value of 1/10 th of the number of individuals. Parameterized uniform crossovers are employed. After two parents are chosen randomly from the full old population (including chromosomes copied to the next generation in the elitist pass), at each gene a biased coin is tossed to select which parent will contribute the offspring. To prevent premature convergence of the population, at each generation one or more new members of the population are randomly generated from the same distribution as the original population. This process has the same effect as applying at each generation the traditional mutation process with a small probability. As it was commented above, all chromosomes of the first generation are randomly generated. This scheme was based on [14]. Figure 3 depicts the evolutionary process.

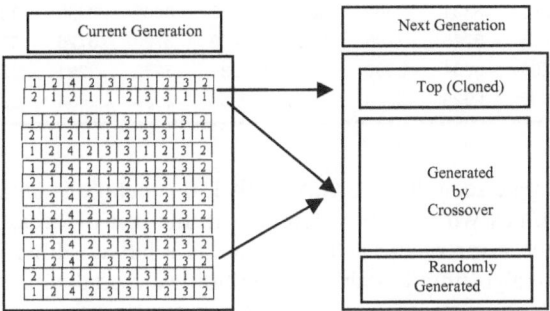

Fig. 3. Evolutionary process used in this approach

4 Numerical Results

The purpose of this section is to show, using a series of numerical examples, how the proposed formulation can be used as an aid to configure a modular component system and to provide an assessment of the performance of the implementation of a genetic algorithm and compare it to simulated annealing algorithm. Nine test problems were used to evaluate the proposed implementation of the G.A. Each experiment considers a problem with 25 figures each one with 2 options of configuration. The module population is comprised by 20 different instances. These 9 problems were first solved through an extensive search, so that their optimal objective functions are known. Each experiment took about 16 hours on a PC with Intel Pentium 4 running at 3.6 GHz under MS windows. Table 1 shows the minimum cost obtained in each case and the results obtained by the S.A. algorithm.

Then the same problems were solved by the proposed G.A. and by the implemented Simulated Annealing. In the case of the G.A., two sets of experiments were run. The first set considered a population size of 30 individuals and 90 iterations. The second set of experiments considered the same population size and 150 generations. All experiments were run 10 times and were totally executed in few seconds in average; the results of the first set of experiments are shown in Table II where optimal solutions are indicated in bold. The results of the second set of experiments are shown in Table 3. A comparison of these two tables shows that, using the above values of the G.A. parameters it was possible to find the optimal solution for 44 (49%) of the 90 tests in the case of the first reported configuration and 82 (91%) over the 90 tests. Table 4 shows the mean deviations (%) between the optimal solution and the best solution obtained from the 10 executions of the proposed G.A.

Table 1. Results obtained with S.A.

	Optimal Solution	1	2	3	4	5	6	7	8	9	10
1	3583	3586	3591	3590	3594	3592	3586	3590	3588	3592	3590
2	3877	3887	3893	3889	3881	3888	3888	3885	3885	3881	3882
3	2622	2623	2637	2636	2634	2630	2634	2636	**2622**	2633	2635
4	3426	3428	3431	3430	3430	3429	3430	3431	3427	2428	3432
5	3491	3505	3501	**3491**	3500	3505	3499	**3491**	3498	3498	3500
6	3290	**3290**	3301	3305	3299	3302	3302	3298	3297	3293	3304
7	3714	3723	3729	3728	3733	3729	3728	3726	3727	3724	3736
8	3279	3283	3284	3280	3283	3281	3282	3281	3280	3280	**3279**
9	2921	2935	2929	2926	2926	2934	2934	2934	2939	2929	2935

for each test. It can be seen that the G.A. with 150 generations obtains the optimum a large number of times with small differences in the rest of occasions. The standard deviation (s) of the error of the different runs is relatively low, indicating that the performance of the algorithm is stable. In addition we used the model and the selected configuration of the Genetic Algorithm to solve a number of randomly generated tests problems. These 12 large problems could not be solved optimally. Therefore, it is an open question how far the best solution found by the G.A. is above the optimum value. The details of each one of the 12 test problems are shown in Table 5. In application like these, exhaustive enumeration would consume CPU time in the order of years. Given the feature of a given test problem, say the problem number 3, the total number of possible combinations corresponds to $10!^{20} = 1,56*10^{31}$. Hence, explicit enumeration is computationally infeasible for this type of problems.

We applied the G.A. 100 times with different random costs values and modular configurations of products in a set of 12 test problems. Computational results are summarized in Table 6. Column "Best" shows the total cost of the best solution found by the genetic algorithm in the 100 runs. Column "% |Best-Ave|/Best" shows the deviation between the best solution and the averaged solution over the 100 runs. Column "1% Opt." shows the percentage of solutions that differed by at most 1% from the best solution found. Column "5% Opt." shows the percentage of solutions that

Table 2. Results from the G.A. with 90 generation

	Optimal Solution	1	2	3	4	5	6	7	8	9	10
1	3583	3586	3585	**3583**	3585	**3583**	3584	3584	3590	3588	**3583**
2	3877	3885	**3877**	**3877**	**3877**	3881	3878	3878	**3877**	**3877**	**3877**
3	2622	2626	2625	2625	2623	2624	2627	**2622**	2631	2624	**2622**
4	3426	3441	3430	3430	3432	3428	**3426**	**3426**	**3426**	3428	**3426**
5	3491	3495	**3491**	**3491**	3492	3503	3495	**3491**	**3491**	**3491**	**3491**
6	3290	**3290**	**3290**	**3290**	3292	3292	3295	3302	**3290**	3291	3291
7	3714	**3714**	**3714**	**3714**	**3714**	3718	**3714**	**3714**	**3714**	**3714**	3718
8	3279	3289	3285	**3279**	3289	3285	3285	**3279**	3290	**3279**	3283
9	2921	**2921**	2930	**2921**	**2921**	**2921**	2926	**2921**	**2921**	**2921**	**2921**

Table 3. Results from the G.A. with 150 generations

	Optimal Solution	1	2	3	4	5	6	7	8	9	10
1	3583	**3583**	**3583**	**3583**	**3583**	**3583**	**3583**	**3583**	**3583**	3585	**3583**
2	3877	**3877**	3880	**3877**	**3877**	3880	3880	**3877**	**3877**	**3877**	**3877**
3	2622	**2622**	**2622**	**2622**	**2622**	**2622**	**2622**	**2622**	**2622**	**2622**	2625
4	3426	**3426**	**3426**	**3426**	**3426**	**3426**	**3426**	**3426**	**3426**	**3426**	**3426**
5	3491	**3491**	**3491**	**3491**	**3491**	**3491**	**3491**	**3491**	**3491**	**3491**	**3491**
6	3290	**3290**	**3290**	**3290**	3294	**3290**	**3290**	3294	**3290**	**3290**	**3290**
7	3714	**3714**	**3714**	**3714**	**3714**	**3714**	**3714**	**3714**	**3714**	**3714**	**3714**
8	3279	**3279**	**3279**	**3279**	3285	**3279**	**3279**	**3279**	**3279**	**3279**	**3279**
9	2921	**2921**	**2921**	**2921**	**2921**	**2921**	**2921**	**2921**	**2921**	**2921**	**2921**

Table 4. Gap among GA, S.A. results (using reference set) and the optimal solutions

Gener.: 90			Gener.: 150			S.A.		
Mean Deviation (%)	St.Dev. of the error	Number of instances	Mean Deviation (%)	St.Dev. of the error	Number of instances	Mean Deviation (%)	St.Dev. of the error	Number of instances
5,86%	6,51%	3	0,56%	1,77%	9	19,26%	7,26%	0
3,61%	6,79%	6	2,32%	3,74%	7	22,96%	9,99%	0
11,06%	10,24%	2	1,14%	3,62%	9	38,14%	20,50%	1
9,63%	13,49%	4	0,00%	0,00%	10	10,51%	4,60%	0
6,02%	11,01%	6	0,00%	0,00%	10	22,34%	13,76%	2
6,99%	11,38%	4	2,43%	5,13%	8	27,66%	14,50%	1
2,15%	4,54%	8	0,00%	0,00%	10	38,50%	10,47%	0
16,16%	13,02%	3	1,83%	5,79%	9	7,01%	4,99%	1
4,79%	10,61%	8	0,00%	0,00%	10	38,00%	14,83%	0

Table 5. Configurations of the 12 test problems

Test	Pieces	Options	Modules
1	20	3	15
2	20	5	15
3	20	10	15
4	30	3	15
5	30	5	15
6	30	10	15
7	40	3	15
8	40	5	15
9	40	10	15
10	50	3	15
11	50	5	15
12	50	10	15

Table 6. Peformance evaluation of the resolution of the 12 test problems

Test	Best	Average	% |Best-Ave|/Best	1% Opt	5% Opt
1	1885	1894,5	0,50	90,00	100,00
2	3658	3696,8	1,06	30,00	100,00
3	4669	4718,2	1,05	60,00	100,00
4	3854	3883,9	0,78	90,00	100,00
5	4880	4913,5	0,69	80,00	100,00
6	6967	7180,8	3,07	10,00	100,00
7	6593	6638,4	0,69	70,00	100,00
8	6302	6354,2	0,83	60,00	100,00
9	14255	14396,3	0,99	50,00	100,00
10	5291	5337,8	0,88	50,00	100,00
11	10322	10419,1	0,94	50,00	100,00
12	12832	12996,8	1,28	30,00	100,00

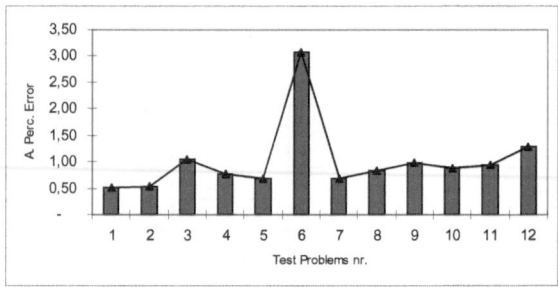

Fig. 4. Average percentual error of the 12 test problems

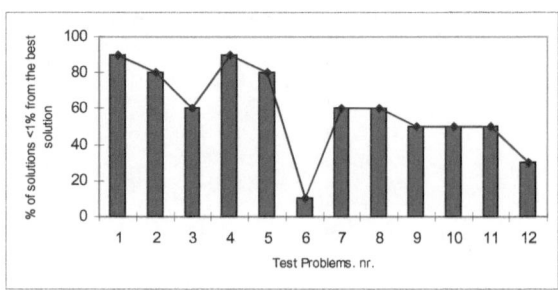

Fig. 5. Percentage of solutions that differed by at most 1% from the best solutions

differed by at most 5% from the best solution found. The G.A. found solutions that were on average always less than 3,07% above the best solution found. For each problem, 55% on average of the solutions were within 1% of optimally. Finally, 100% of the solutions found by the Genetic Algorithm were within the 5% of optimally.

Figure 4 shows the performance of the proposed genetic algorithm applied to the 12 problems in terms of their average percentual error. Figure 5 shows the percentage of solutions that differed by at most 1% from the best solution found for each one of the 12 test problems. As can be observed in the figures 4 and 5, the results obtained are better in those problems where the number of alternatives for each module is lower.

As the number of alternatives increases, the percentage of error also increases and the number of times that obtained solutions that differed by at most 1% from the best solution obtained decreases.

5 Conclusions

A genetic algorithm was presented for selection and optimization in transforming structures into modular ones. This problem is found in industries that face gradual or complete transformation of these products to modular based assemblies. The optimization problem considers the cost involved in using a set of modules to obtain a number of modular assemblies that could substitute a set of non-modular products or structures. A set of problems was generated, and their optimal solutions were obtained with exhaustive search. The GA was tested and the results show that the GA obtained the optimal solution in all the cases maintaining a low variability of the results. The proposed algorithm is computationally feasible for large problems. It finds good solutions with moderate CPU times (assuming that the best solutions found are close to optimum). The proposed formulation is general in the sense that products can have any number of modules. Through modularity, the number of different parts to be purchased for an assembled product set may be significantly reduced while achieving a sufficient variety by combination of different modules.

References

1. Meng, X., Jiang, Z., Huang, G.Q.: On the module identification for product family development. Int. J. Adv. Manuf. Technol. 35, 26–40 (2007)
2. Baldwin, C.Y., Clark, K.B.: Managing in an Age of Modularity. Harvard Business Review 75(5), 84–93 (1997)
3. Pandremenos, J., Paralizas, K., Salonitis, K., Chryssolouris, G.: Modularity concepts for the automotive industry: A critycal review. CIRP Journal of Manufacturing Science and Technology 1, 148–152 (2009)
4. O'Grady, P., Liang, W.-Y.: An internet-based search formalism for design with modules. Comp. Ind. Eng. 35(1–2), 13–16 (1998)
5. Bi, Z.M., Zhang, W.J.: Modularity Technology in Manufacturing: Taxonomy and Issues. Int. J. Adv. Manuf. Technol. 18, 381–390 (2001)
6. Asan, U., Polat, S., Serdar, S.: An integrated method for designing modular products. Journal of Mfg. Tech. Mgmt. 15(1), 29–49 (2004)
7. Hornby, G.S., Lipson, H., Pollack, J.B.: Generative representations for the automated design of modular physical robots. IEEE Transactions on Robotics and Automation 19, 703–719 (2003)
8. Babu, B.S., Valli, P.M., Kumar, A.V.V.A., Rao, D.N.: Automatic modular fixture generation in computer-aided process planning systems. Proceedings of The Institution of Mechanical Engineers Part C-Journal of Mechanical Engineering Science 219(10), 1147–1152 (2005)

9. Retik, A., Warszawski, A.: Automated design of prefabricated building. Building and Environment 29(4), 421–436 (1994)

10. Kingston, J.: Modularity as an Enabler for a More Efficient Commercial Small Satellite Program. In: Proceedings of the 17th Annual AIAA/USU Conference on Small Satellites, SSC03-III-8 (2003)

11. Ishii, K.: Product Modularity: A Key Concept in Life-Cycle Design. In: Frontiers of Engineering: Reports on Leading Edge Engineering from the 1996 NAE Symposium on Frontiers of Engineering (1997)

12. Holland, J.H.: Adaptation in Natural and Artificial Systems. University of Michigan Press, Ann Arbor (1975)

13. Metropolis, N., Rosenbluth, A., Rosenbluth, M., Teller, A., Teller, E.: Equation of State Calculations by Fast Computing Machines. J. Chem. Phys. 21(6), 1087–1092 (1953)

14. Goncalves, J.F., Resende, M.G.C.: An evolutionary algorithm for manufacturing cell formation. Comp. Ind. Eng. 47, 247–273 (2004)

TDoA Based UAV Localization Using Dual-EKF Algorithm

S.C. Lee, W.R. Lee, and K.H. You

Sungkyunkwan University, Suwon, 440-746, Korea
lsch1977@lycos.co.kr, wooram@ece.skku.ac.kr,
khyou@ece.skku.ac.kr

Abstract. Most UAV(unmaned aerial vehicle) systems use GPS signals only to locate the emitter's position. However GPS signals contain unwanted information contaminated by environmental components and many interference signals. In this paper, to obtain TDoA signal, we use two UAVs which are equipped with embedded wireless sensors. Under the real geolocation circumstance, it is very difficult to estimate the emitter's position exactly due to environmental noise. In this paper we use the dual-EKF algorithm to obtain the optimal estimation of state values and unknown parameters. The dual-EKF algorithm overcomes the weakness of EKF algorithm which has been widely used in geolocation problem. The performance of our proposed algorithm will be demonstrated through some simulations of UAVs.

Keywords: Time difference of arrival, unmaned aerial vehicles, dual-EKF, localization.

1 Introduction

The tracking techniques for locating emitter's position in UAV are studied in many approaches these days. The localization technique includes many applications such as robot SLAM, search for missing child and rescue against disaster.[1-2] There are two main methods for acquiring time difference of arrival(TDoA) signal. The first method measures time of arrival(ToA) from each UAV to emitter. Using the ToA values, the TDoA can be represented in the form of time difference. The second method obtains TDoA signals using the cross-correlation of each received signal.

Most techniques that use TDoA require at least three UAVs. However recently there has been a trend towards reduced cost. The UAV needs only very basic sensors which measure the time of arrival of signals at each receiver and can obtain TDoA signal between UAV and emitter. If there is no measurement noise and if there are many adjacent UAVs, the position estimation is easily obtained by hyperbola method. However, there exist unwanted noise components in real circumstances. In this case, the hyperbolic curves will no longer intersect at one precise position. This problem can be solved with least squares method [3-4] calculating all measurement signals. Fang[5] has studied closed-form of hyperbolic fix through augmented TDoA measurements. However Fang's method cannot be applied to overdetermined situation. Moreover this method needs more TDoA signals.

D. Ślęzak et al. (Eds.): CA 2009, CCIS 65, pp. 47–54, 2009.

In tracking algorithm with TDoA signals there are nonlinearity components caused by noises.[6] In the geolocation applications, the extended Kalman filter(EKF) algorithm has been widely used.[7] However EKF has the divergence problem in the calculation process and poor performance in serious noise circumstance. In this paper, we propose dual-EKF algorithm which can obtain optimal estimation of state and system's unknown parameters. Finally some simulation results demonstrate the performance of dual-EKF under the circumstance in which two TDoA signals received from two UAVs are used.

2   Hyperbolic Curve Method for Emitter Localization

2.1   Analytical Methods

In this section we show a general hyperbolic curve method which has been widely used under the ideal circumstance. If there exists more than two UAVs, we can receive more associated TDoA signal values and obtain more precise position of unknown emitter. The emitter position is located as the intersection of hyperbolic curves. Figure 1 represents the general hyperbolic curves under the ideal circumstance. Here we consider only three UAVs for simplicity of method.

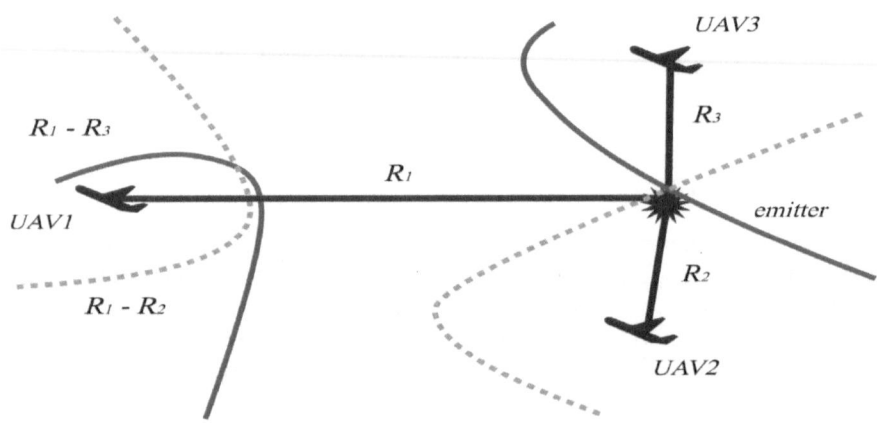

Fig. 1. Geometric method using hyperbola

We assume that emitter lying on position $r^0 = (x, y)$ sends out wireless signals to UAVs during the discrete sampling time. Position of three UAVs are supposed to be known as $r_i = (x_i, y_i), i = \{0, 1, 2\}$, respectively and each UAV receives wireless signals from emitter at each discrete sampling time.

$$t_i^0 = \frac{1}{c} r_i^o$$

$$r_i^0 = \sqrt{(x_i - x)^2 + (y_i - y)^2} \tag{1}$$

where r_i^0 means the distance from emitter to UAVs and c represents the propagation speed. Using the norm of a distance, we can derive the following equation.

$$r_i^o = \|r_i - r^o\|$$
$$r_{i1}^o = ct_{i1} = r_i^0 - r_1^0 \tag{2}$$

where t_{i1} represents TDoA signal between the i-th receiver and the first one.

In this paper, we use only two receivers(UAVs) for algorithm's simplicity. TDoA signals using two UAVs are represented as follows.

$$TDoA = \frac{1}{c}\left(\|r_2 - r^o\| - \|r_1 - r^o\| + v_{21}\right) \tag{3}$$

where v_{21} means zero mean Gaussian noise under the real circumstance case and r_2, r_1 mean the position of UAV2, UAV1, respectively.

2.2 System Modeling

We assume that the moving emitter with constant velocity sends wireless signals and the two UAVs with elliptical orbit movement can obtain the wireless signal from emitter. Through this process, we obtain TDoA measurement values. However TDoA signals contain unwanted nonlinear error components. As a robust solution, the dual-EKF approach using TDoA signals can be more efficient among geolocation methods. To use the dual-EKF algorithm, we need the state equation using the emitter's position and velocity as state variables. We set up the following state equation which represents two dimensional position and time interval.

$$s(k+1) = As(k) + Bu(k) + w(k)$$

$$A = \begin{bmatrix} 1 & 0 & 1 & 0 \\ 0 & 1 & 0 & 1 \\ 0 & 0 & 0 & 0 \\ 0 & 0 & 0 & 0 \end{bmatrix}, \quad B = \begin{bmatrix} 0 & 0 \\ 0 & 0 \\ T & 0 \\ 0 & T \end{bmatrix} \tag{4}$$

where $s(k) = [x \; y \; \Delta x \; \Delta y]^T$, (x, y) represents the emitter's position, $(\Delta x, \Delta y)$ means the positional variation at each sampling time, $u(k) = \begin{bmatrix} v_x & v_y \end{bmatrix}^T$ means the known velocity of a moving emitter, T is sampling time, and $w(k)$ means the process noise as additive white Gaussian noise(AWGN). The output equation can be expressed by the measurement of TDoA value as

$$z(k) = h(s^0(k), v(k))$$
$$= \frac{1}{c}\left(\left\| s^0(k) - r_1 \right\| - \left\| s^0(k) - r_2 \right\|\right) + v(k) \tag{5}$$

where $s^0(k)$ means the emitter's position and $v(k)$ is the measurement noise as AWGN.

2.3 Dual-Extended Kalman Filter Algorithm

In this section, to apply geolocation problem, we propose dual-EKF algorithm whose plant model is augmented by weight filter(e.g., neural network). The dual-EKF algorithm combines the Kalman state and weight filter. This algorithm can estimate both the state and model parameter through only measurement values. The major process is that two EKF algorithms cooperate concurrently at every time-step. The current model parameter $\hat{w}_k$ help the EKF state-filter to estimate state values while the EKF weight-filter updates the weight parameters with the current state estimate $\hat{x}_k$.[8] During the update process, the weight filter is used as weight training method in neural network. The weight filter has the high convergence than back-propagation in neural network.

To apply the dual-EKF algorithm, we initialize the estimation of state and weight values as

$$\hat{w}_0 = E[w], \quad P_{w_0} = E\left[(w - \hat{w}_0)(w - \hat{w}_0)^T\right]$$
$$\hat{x}_0 = E[x_0], \quad P_{x_0} = E\left[(x_0 - \hat{x}_0)(x_0 - \hat{x}_0)^T\right] \tag{6}$$

The time update equations for the weight filter and the state filter are

$$\hat{w}_k^- = \hat{w}_{k-1}$$
$$P_{w_k}^- = P_{w_{k-1}} + R_k^r$$
$$\hat{x}_k^- = F(\hat{x}_{k-1}, u_k, \hat{w}_k^-)$$
$$P_{x_k}^- = A_{k-1}P_{x_{k-1}}A_{k-1}^T + R^v \tag{7}$$

In equation (7), w corresponds to values of unknown parameters. The weight parameters(w) are used in the design process of dynamics system. The innovations covariance(R_k^r) is set to an arbitrary diagonal matrix in which the elements are close to zeros for weight training. R_k^r affects the tracking performance and convergence rate. The measurement update equations for the state filter and the weight filter are

$$K_k^x = P_{x_k}^- C^T (C P_{x_k}^- C^T + R^n)^{-1}$$

$$\hat{x}_k = \hat{x}_k^- + K_k^x (y_k - C\hat{x}_k^-)$$

$$P_{x_k} = (I - K_k^x C) P_{x_k}^-$$

$$K_k^w = P_{w_k}^- (C_k^w)^T (C_k^w P_{w_k}^- (C_k^w)^T + R^e)^{-1}$$

$$\hat{w}_k = \hat{w}_k^- + K_k^w (y_k - C\hat{x}_k^-)$$

$$P_{w_k} = \left(I - K_k^w C_k^w\right) P_{w_k}^-$$ (8)

$$C = \frac{\partial z}{\partial x}, C_k^w = \frac{\partial z}{\partial w}$$

The noise covariance (R^e) is a constant diagonal matrix and can be set arbitrarily as $0.5I$. The matrix of I means an identity matrix. The first order linearization of measurement signal is necessary in order to approximate the nonlinear dynamics. The system applied to geolocation problem is shown schematically in figure 2.

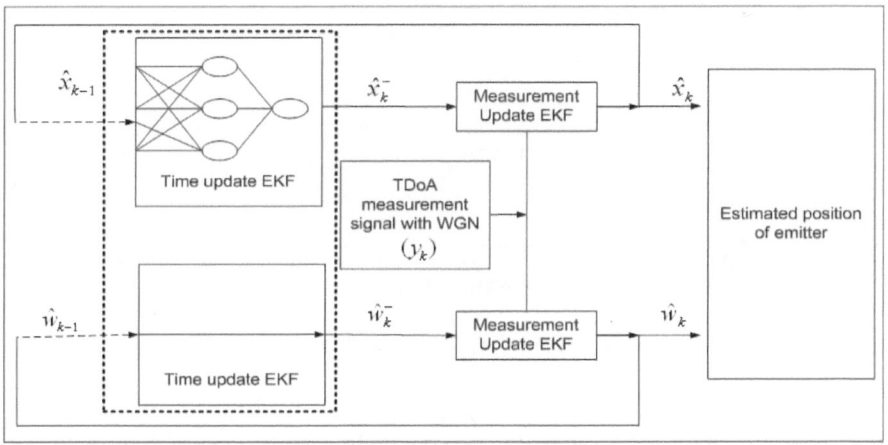

Fig. 2. Process of dual-EKF algorithm in UAV localization

3 Simulation Results

In this section, we demonstrates the effectiveness of the proposed geolocation method with the dual-EKF algorithm through some simulations. In figure 3, we suppose that the emitter moves with a constant speed except that emitter changes its direction every 19 sec. The two UAVs are initially located at the position of (8,6) km and (6,11) km with constant altitude and move following the formation of circular orbits (radius of 1 km). In the simulation, we can obtain 500 TDoA measurement values to apply the dual-EKF algorithm. The initial estimate of emitter location is randomly obtained from a white Gaussian distribution in the ideal emitter. The noise from TDoA signals has the Gaussian distribution of 0.0583(standard deviation). The standard deviation of process noise is supposed to be 0.0591.

In Fig. 3, the dotted line means the real movement of the emitter while the solid line represents the estimated trajectory using the dual-EKF algorithm. As shown in figure 3, there are some differences between the real path and dual-EKF estimation around the range of 5 km and 6.5 km in the x-axis. However, after that, the trained weight filter enables the state values to track the real path more accurately. Therefore, we can obtain the estimated path which is similar to real path with updated weight.

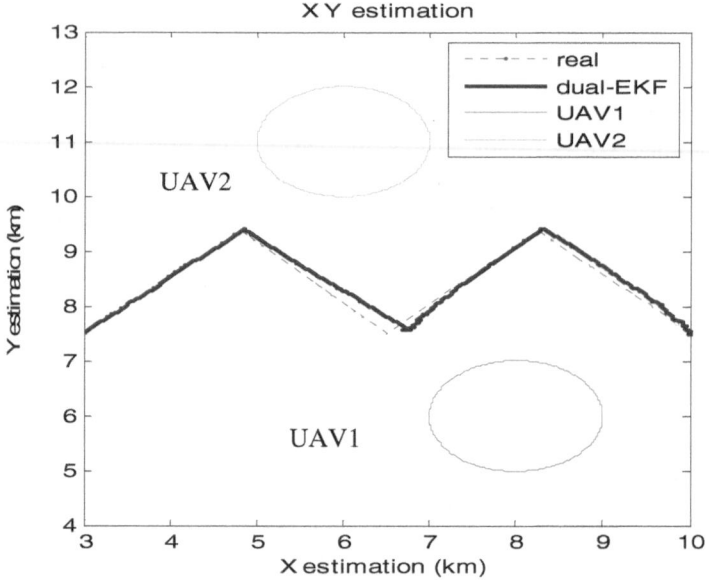

Fig. 3. Trajectory estimation of real path

Figure 4 shows the performance of the proposed localization algorithm with different path of emitter. The dotted line is the emitter's trajectory and the solid line is the one by dual-EKF with the weight filter and state filter. The weight value of the unknown parameter is updated on each estimation step.

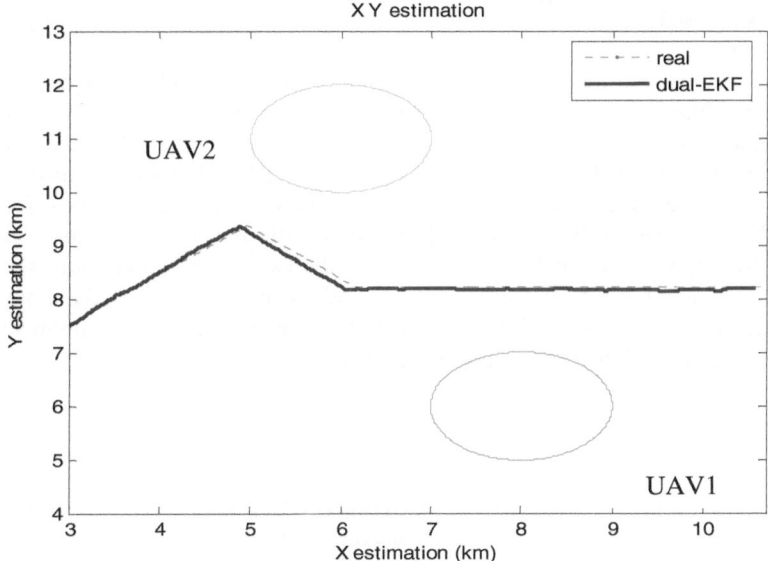

Fig. 4. Trajectory estimation for different path

Figure 5 represents the norm of position error in figure 4, i.e., $\left\| s^0(k) - \hat{s}^0(k) \right\|$. The general geolocation method using GPS signals has the limit of 100 m error. In this paper, we supposed that the emitter moves with constant velocity(105 *m/s*). As shown in Fig. 5, in spite of emitter's fast movement and large unexpected noise, the estimated error values through dual-EKF algorithm is limited within the boundary of ±100 *m* positional error. The effectiveness of algorithm is confirmed and the estimated trajectory is shown to be close to real path.

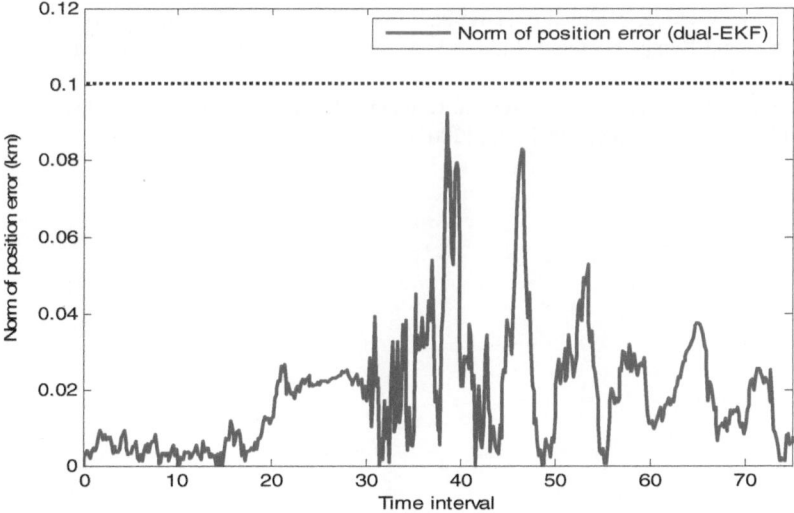

Fig. 5. Norm of position error

4 Conclusion

In this paper, we introduced a geolocation algorithm using TDoA signals from two UAVs. The dual-EKF method can estimate state value and unknown parameter value of dynamic state model with only measurement TDoA values. In case of position tracking system with TDoA signals, it requires much calculation efforts from the linearization process. Through the simulation results we demonstrated the effectiveness of dual-EKF algorithm. It is confirmed that the position estimation using dual-EKF is close to the trajectory of real emitter. Through the two UAVs simulation, we verified that the positional error of the estimation values resides within the boundary of $\pm100m$ error. Furthermore, with only two UAVs for the TDoA measurement values, we realized geolocation algorithm with dual-EKF.

References

1. Okello, N., Musicki, D.: Emitter Geolocation with two UAVs. In: Proceedings of Information, Decision & Control Conf., Adelaide, Australia (February 2007)
2. Fletcher, F., Ristic, B.: Recursive estimation of Emitter location using TDoA measurements from two UAVs. In: 10th International Conf. Information Fusion, July 2007, pp. 1–8 (2007)
3. Drake, S., Dogancay, K.: Geolocation by Time difference of arrival using hyperbolic asymptotes. In: IEEE International Conf. Acoustics, Speech & Signal Processing 2004 Proceedings, May 2004, vol. 2, pp. 361–364 (2004)
4. Hashemi-Sakhtsari, A., Dogancay, K.: Recursive least squares solution to Source tracking using Time difference of arrival. In: IEEE International Conf. Acoustics, Speech & Signal Processing 2004 Proceedings, May 2004, vol. 2, pp. 385–388 (2004)
5. Fang, B.T.: Simple solutions for hyperbolic and related position fixes. IEEE Trans. Aerospace & Electronic Systems 26(5), 748–753
6. Ho, K.H., Chan, Y.T.: Solution and Performance analysis of geolocation by TDoA. IEEE Trans. Aerospace & Electronic Systems 29(4), 1311–1322 (1993)
7. Najar, M., Vidal, J.: Kalman tracking based on TDoA for UMTS mobile location. In: IEEE International Symp. Personal, Indoor & Mobile Radio Communications, vol. 1, pp. B45–B49 (2001)
8. Eric Wan, A., Nelson, A. T.: Nelson.:Kalman filtering and neural networks, pp. 123–133. John Wiley & Sons, Inc, Chichester (2001)

A Design of Framework for Smart Services of Robots in Intelligent Environment

Joo-Hyung Kim[1], Dong Won Kim[2,*], Bum-Jae You[3], and Gwi-Tae Park[1,*]

[1] Department of Electrical Engineering, Korea University, Seoul, Korea
{proteus99,gtpark}@korea.ac.kr
[2] Department of Digital Electronics, Inha Technical College, Incheon, Rep. of Korea
dwnkim@inhatc.ac.kr
[3] Center for Cognitive Robotics Research, Korea Institute of Science and Technology, Seoul, Korea
ybj@kist.re.kr

Abstract. This paper presents a framework for smart services of robots in an intelligent environment. Within such intelligent environment, the target platform consists of service robots and a framework for smart services including task allocation and task scheduling. For task allocation, we use an auction-based method and a knapsack problem algorithm. In this paper, the characteristics of the algorithm for allocation balancing are to delegate executing task to another robot, to reallocate the delayed task to more than one robot, and to withdraw over-allocated robots from the current task.

Keywords: intelligent environment, smart services, task management.

1 Introduction

Recently, computer networks have developed by many researchers and have become an important part of our daily lives. Intelligent environment is one of such computer networks, and it enables to accomplish diverse services, such as visitor guidance and indoor surveillance [1]. The environment is able to monitor what is happening in it, to communicate with their inhabitants, and to make a decision something important. In [2], sensing and reacting context, information sensed to characterize the situation of the people, activities, interaction between user and application are prominent characteristic of the intelligent environment. In [3], "Intelligent Space" was proposed by Lee *et al* for interaction between human and space based on intelligent environment. In the space, a robot is used as a physical agent for offering human-centered services [4], [5].

In such an intelligent environment, the more people require diverse services, the more complex and huge the system becomes, and the role of robots as physical agent for offering smart services becomes more and more various and important. Thus, we need network based service framework to manage service tasks of robots.

D. Ślęzak et al. (Eds.): CA 2009, CCIS 65, pp. 55–61, 2009.
© Springer-Verlag Berlin Heidelberg 2009

2   Architecture for Intelligent Environments

In this paper, we use a Resource Sharing Architecture (RSA) [6] as an architecture for an intelligent environment. The main focus of RSA is to share physical resources, and to organize them effectively for supporting high quality services. And, the most important characteristic of RSA is that it makes a robot overcome something physical. The structure of RSA consists of physical resources and service objects. Physical resources are devices such as mobile robots, cameras and so on. Some of them obtain data from the environment, and transmit the data to service objects via local area network. The others provide information to human who wants to know. Whereas service objects, such as robot navigation and room cleaning service, create significant information using data transmitted from physical resources and inform the information to other services, so that an intelligent space can obtain states and abilities of physical devices connected with network. According to the relation of each service object, they are classified into three categories as follows:

1. Fundamental service: It's a service object directly connected with physical devices. Such objects obtain data from sensors, or transmit useful information to human using devices. And the services inform data to inherited service objects.
2. Inherited service: It's organized by more than one fundamental service object and other inherited service objects. It uses the results of fundamental services, and creates visual or acoustic information for transmitting to iSpace service.
3. iSpace service: It's an intelligent service object by combined diverse service objects as mentioned above (i.e. fundamental service and inherited service), which is able to carry out an intelligent task such as room cleaning or visitor guidance.

3   Framework for Smart Services Based on RSA

In this paper, we design a multi agents based framework using a RSA. The framework consists of some robot groups and a task management group as shown in Fig. 1. In a robot group, each robot agent uses some services to control each robot. In a task management group, each task management agent can know what the space should do using information obtained from the diverse services.

3.1   Robot Agent

A robot agent monitors the state of a robot, and creates services to execute the assigned tasks. When a task is assigned to a robot agent, a robot starts carrying out the task using created services. If the executing task is stopped by some event, the agent informs the state of the interrupted task to a task management agent so as to be continued the task by other robot agents.

3.2   Robot Group

According to the ability of each robot, they can be divided into more than one group. In case of a cooperative work, a task management agent makes a robot group, and add robots for carrying out the assigned task to the group.

3.3 Task Management Agent

A task management agent plans the schedule of the robot's task. To manage the tasks, the agent acquires information from some services and robot agents, and creates the task list of robots to manage assigned tasks, and allocates the tasks to best candidates. If interrupted tasks remain in the task list, the agent can reallocate the tasks to other robots. All processes related to task scheduling and task allocation are decided by the task management algorithm which is included in the agent. We will deal with the algorithm for task management in the following section.

3.4 Task management group

A task management group is a set of task management agents. If the group coordinates the needed service to an appropriate task management agent, then each task management agent executes smart services.

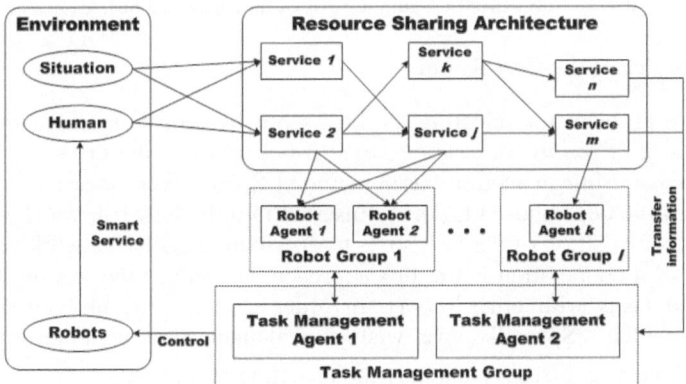

Fig. 1. Overview of framework for smart services

4 Smart Service Management Algorithm for Multi-robot

In Intelligent environment, people can be provided physical service by robots and other devices. Thus, a task management agent should manage the tasks of both robots and devices. In this paper, we only focus on the task of robots, and the assumptions of tasks that we consider are described as follows:

1. A task can be decomposed into more than one independent sub-task.
2. A task can be interrupted by an event, and the task can be continued by other robots.

Fig. 2 depicts a flow of an algorithm for task management. A task management agent selects robots to carry out the assigned tasks, and informs to a robot agent. The characteristics of the proposed task management algorithm are to delegate executing task to another robot, to reallocate a delayed task to more than one robot and to

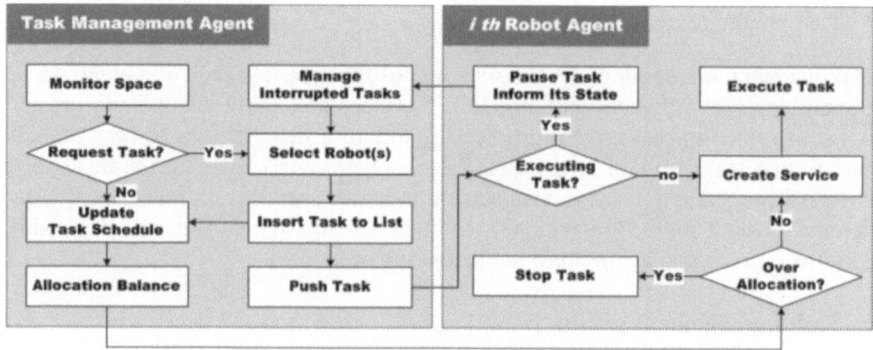

Fig. 2. Flow of task management

withdraw over-allocated robots from the current task. In this paper, we implement two algorithms for task management of service robot as followed subsections.

4.1 How to Select the Best Robot

To solve the problem for selecting the best robot, we consider *single-robot of tasks* and *single-task of robots*. In case of *single-robot of tasks*, the tasks are just inserted into the task list using an auction-based method [7]. However, in case of *single-task of robots*, when the decomposed tasks are inserted into the task list, we should consider effective decomposition of the task so as to distribute equally the workload of robots. In the case of a cooperation work, this approach can reduce the waiting time to start the assigned tasks with other robots. In order to solve the problem, we apply a knapsack problem. Specially, we wish to calculate the workload $W[n,\overline{w}]$ using equation (1), so as to distribute equally the task time of all robots.

$$W[n,\overline{w}] = \begin{cases} \max(W[n-1,\overline{w}], w_n + W[n-1,\overline{w}-w_n]) & (w_n \le \overline{w}) \\ W[n-1,\overline{w}] & (w_n > \overline{w}) \end{cases} \quad (1)$$

where $\overline{w}$ is the average time of tasks, w_i is the expected time of ith task, and n is the number of robots. Here, we calculate all the values of the array $W[n,\overline{w}]$ using the recursive expressions above to calculate subsequent values.

4.2 Allocation Balancing

In general, insufficient allocation of physical resources may bring about the task delay. On the other side, over-allocation for completing an assigned task quickly is not good. A good allocation is to complete an assigned task almost at a prearranged time and to carry out tasks as many as possible using given robots. In this paper, we consider the deadline of assigned task for dynamic task reallocation. The objective of allocation balancing is to reallocate a delayed task to other robots and to withdraw over-allocated robot from the current task. The algorithm predicts the reallocation point x according to the progress of the assigned task. The equation for predicting is expressed as follows:

$$T = \sum_{i=1}^{x} \left[n_i \quad t_i'/c \right] + \sum_{i=x+1}^{l} \left[n_i \quad t_i'/(c+p) \right] \tag{2}$$

where l is the number of total task, n_i is ith task, t_i' is the measured time of ith task, c is the number of allocated robots, p is the number of predicted robot for additional allocation, and x is prediction step. T includes the number of total task and the time of total task.

5   Case Study: Simulation of Smart Services of Multi-robot

In this paper, we evaluate the proposed algorithm through two simulation scenarios. In the simulations, there are three mobile robots. And, we set up cleaning scenario to evaluate the algorithm, and assume that the robots have same ability. In the first scenario, we compare an auction-based method with a knapsack problem to evaluate task allocation algorithm. In second scenario, we evaluate the proposed algorithm for allocation balancing according to a change of the task time.

5.1   Evaluation for Selecting the Most Appropriate Robots

The objective of this simulation is to evaluate the proposed task management algorithm using a knapsack problem whether the algorithm distributes equally the workload of robots. The scenario for evaluating is as follows:

Scenario 1
Step 1: *There are three robots. And 30 tasks will be allocated to robots. The task time is 10 or 50. Task time is generated randomly every step.*   ***Step 2***: *By an auction-based method, a task management agent allocates 30 tasks to robots.*   ***Step 3***: *By a knapsack problem, a task management agent allocates the same tasks to robots.*

Fig. 3 represents the results of task allocation using an auction-based method and using a knapsack problem. The lines in the figure show the cumulative time of tasks assigned to robots every step. In the experimental result, our method for the balance of robot's workload shows more effective than an auction-based method.

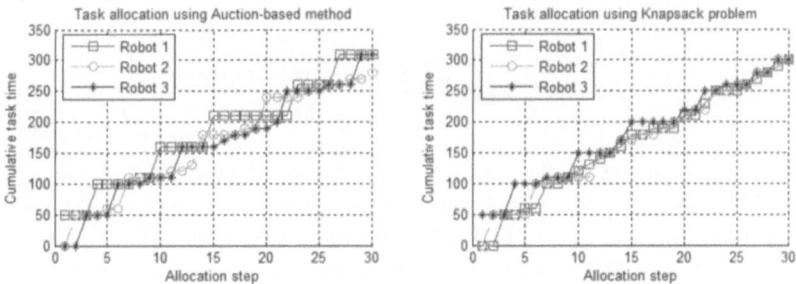

Fig. 3. Comparing auction-based method with knapsack problem

5.2   Evaluation for Allocation Balancing

In this scenario, we evaluate the proposed algorithm which allocates robots dynamically to complete the assigned task by deadline. The scenario is as follows:

Scenario 2
Step 1: *Robot 1 and Robot 2 have to carry out Task B. However, the executing time of Task B is reduced because of the delay of Task A.*
Step 2: *According to the task schedule, a task management agent allocates Task B to Robot 1 and Robot 2.*
Step 3: *A task management agent monitors the state of Task B, and allocates the task to Robot 3 so as to complete the task within deadline.*
Step 4: *In opposition, when Task A is completed earlier than the schedule, a task management agent withdraws over-allocated robots from task B.*

At first, we measured the average time (i.e., about 43,000 seconds) of Task B, and considered two cases which are the increase of the task executing time and the decrease of the time. Table 2 and Table 3 depict the result of the measured task time and the predicted point for additional allocation according to the change of the task time. In the tables, the value of 49th / 50th means that the 49th task is allocated to Robot 3 in 50 independent tasks. And, we calculated the ratio of task delay from the result of the tables using equation (3).

Table 1. Case 1: Decrease of the executing time of the task

Task time	42,000	41,000	40,000	39,000	38,000
Measured time	42,182	41,588	40,994	39,504	37,857
Predicted point	49th / 50th	46th / 50th	43th / 50th	37th / 50th	36th / 50th

Table 2. Case 2: Increase of the executing time of the task

Task time	44,000	45,000	46,000	47,000	48,000
Measured time	43,897	44,038	45,859	47,143	48,229
Predicted point	47th / 50th	46th / 50th	44th / 50th	42th / 50th	41th / 50th

$$\text{Ratio of delay}\,(\%) = \frac{task\ time - measured\ time}{task\ time} \tag{3}$$

In the experiment result, the ratio of task delay is about 1.054% in the case 1 and about -0.379% in the case 2. Fig. 4 shows the reliability of the proposed algorithm for allocation balancing. In case 1, Task B is delayed in an average of 425 seconds with a standard deviation of 487 seconds, and the reliability is about 82%. In case 2, Task B is completed early in an average of 167 seconds with a standard deviation of 458 seconds, and the reliability is about 87%.

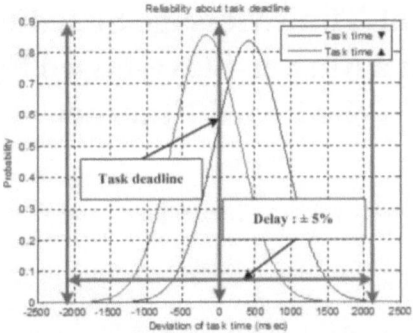

Fig. 4. Reliability of task deadline

6   Conclusions and Future Works

In this paper, we proposed a framework for smart services of robots in an intelligent environment. The important characteristics of the proposed framework are to delegate executing task to a robot, to reallocate the delayed task to more than one robot and to withdraw over-allocated robots from the current task. In near future, we improve the proposed framework for carrying out complex services using both robots and physical devices.

Acknowledgments. This work was supported by the Korea Institute of Science and Technology(KIST).

References

1. Weiser, M.: The computer for the twenty-first century. Sci. Amer. 265, 94–100 (1991)
2. Qin, W., Suo, Y., Shi, Y.: CAMPS: A Middleware for Providing Context-Aware Services for Smart Space. In: Chung, Y.-C., Moreira, J.E. (eds.) GPC 2006. LNCS, vol. 3947, pp. 644–653. Springer, Heidelberg (2006)
3. Lee, J.H., Hashimoto, H.: Intelligent Space: its concept and contents. Adv. Robotics Journal 16(4), 265–280 (2002)
4. Yamaguchi, T., Sato, E., Takama, Y.: Intelligent Space and Human Centered Robotics. IEEE Trans. on Industrial Electronics 50(5) (October 2003)
5. Lee, J.H., Morioka, K., Ando, N., Hashimoto, H.: Human Centered Ubiquitous Display in Intelligent Space. In: 33rd Annual Conference of Industrial Electronics Society 2007, Taipei, Taiwan, November 2007, pp. 22–27 (2007)
6. Lee, B.J., Lee, H.G., Lee, J.H., Park, G.T.: A design of a data accessing service for a real-time vision service in the resource sharing architecture. In: Proc. IEEE Intl. Conf. Robotics Autom., pp. 1235–1240 (2002)
7. Shah, K., Meng, Y.: Communication-Efficient Dynamic Task Scheduling for Heterogeneous Multi-Robot Systems. In: Proc. IEEE Int. symposium Comp. Intelli. Robotics & Autom., pp. 230–235 (2007)

EDF-Based Real-Time Message Scheduling of Periodic Messages on a Master-Slave-Based Synchronized Switched Ethernet

Myung-Kyun Kim[1], Liang Shan[1], and Wang Yu[2]

[1] University of Ulsan, Mugeo-Dong, Nam-Gu, Ulsan 680749, Korea
mkkim@ulsan.ac.kr
[2] University of North Carolina at Charlotte, 9201 University City Blvd
Charlotte, NC 28223-0001, USA
yu.wang@uncc.edu

Abstract. Switched Ethernet has many features for real-time communications but cannot guarantee the timely delivery of a real-time message due to possible collisions on the output ports. This paper first suggests a feasible condition for real-time communication of periodic messages on a master-slave-based synchronized switched Ethernet. Then an EDF (Earliest Deadline First)-based scheduling algorithm that satisfies the proposed scheduling condition is proposed. The master node checks the feasible condition for messages and makes a message transmission schedule for feasible messages. The proposed scheduling algorithm can handle dynamic message requests and performs the real-time communication without any modification in the switch. The performance of the proposed scheduling algorithm has been evaluated by simulation to show the timely delivery of real-time messages and the real-time communication capacity of the switched Ethernet.

Keywords: Real-time communications, switched Ethernet, scheduling algorithm, EDF-based message scheduling.

1 Introduction

Ethernet has become the most widely used network technology for data communication in office environments. Ethernet is very easy to install and shows good performance when the number of nodes on the network is small. However, due to sharing of a communication bus among all nodes, the performance decreases rapidly if the number of nodes on the network increases. On the other hand, switched Ethernet eliminates messages collision on the network by traffic isolation of messages to each port, which greatly enhances the performance of the switched Ethernet. Switched Ethernet has many attractive features for real-time communications, but cannot guarantee message transmission within the deadline mainly due to the collision on the output port.

Many research efforts have been done in order to use the switched Ethernet as a real-time communication network in the industrial environment. George et al. [3] analyzed the architectures of switched Ethernet networks and presented a method to

D. Ślęzak et al. (Eds.): CA 2009, CCIS 65, pp. 62–70, 2009.

minimize end-to-end delays by using network calculus theory. Lee et. al. [5] analyzed the real-time performance of the switched Ethernet and showed that can be used as an industrial communication network. Varadarajan and Chiueh [8] have proposed a real-time communication protocol called EtheReal. Their approach, however, has no support for hard real-time communication and no explicit support for periodic messages which is required for industrial applications. Another work by Loser and Hartig [6] and Mifdaoui et. al. [1] used a traffic shaper to regulate the message traffic entering the switched Ethernet and to bound end-to-end delay. Their method, however, cannot guarantee the timely delivery of the messages per message basis. Hoang et. al. [4] attempted to support real-time communication of switched Ethernet by adding the real-time layer in both end nodes and the switches. Instead of using FIFO queuing, packets are queued in the order of deadline in the switch. A source node, before sending a real-time periodic message to the destination node, has to establish a real-time channel that is controlled by the switch for guaranteeing the deadline. Almeida et al. [2] have proposed FTT-CAN protocol which adds flexibility to the real-time system based on CAN and have extended the FTT paradigm to the Ethernet to support real-time communication on the synchronized Ethernet [7].

This paper first analyzes the feasible condition for real-time communication on a switched Ethernet. The feasibility of real-time messages on the Ethernet was analyzed by Pedreiras et al. [7], but, to the best of our knowledge, there has been no research result on the feasibility of real-time messages on a switched Ethernet until now. A message scheduling algorithm that satisfies the suggested feasibility condition is also proposed in this paper. We implemented the proposed scheduling algorithm and evaluated the performance of our algorithm by simulation to show the timely delivery of real-time messages and the real-time communication capacity of the switched Ethernet. The simulation results show that all of the messages which satisfy the feasible condition have been transmitted within their deadlines. The simulation result also shows that, in the case of Ethernet, the number of scheduled messages on the network is almost constant according to the number of nodes on the network. On the contrary, in the case of the switched Ethernet, the number of feasible messages was increased linearly according to the number of nodes on the network.

The rest of paper is organized as follow. The master-slave-based synchronized switched Ethernet model and message transmission model on the switched Ethernet are described in Section 2. In Section 3, the feasible condition of real-time messages and an EDF-based real-time message scheduling algorithm that satisfies the condition are described. Section 4 presents the performance evaluation of the proposed scheduling algorithm and Section 5 concludes the paper.

2 Message Transmission Model on the Synchronized Switched Ethernet

In this section, a message transmission model on the switched Ethernet for the proposed hard real-time communication is described. The switched Ethernet consists of one master node and many slave nodes as shown in Fig. 1. The switched Ethernet operates in full-duplex mode, where each node is connected to the switch through a pair of links which operate independently: a transmission link and a reception link.

The master node handles the real-time communication of periodic messages. When a slave node has a new periodic message to send, it transmits the real-time specification of the periodic message to the master node, and the master checks the scheduling condition for the periodic message and makes a feasible schedule for the transmission of the message if it is schedulable.

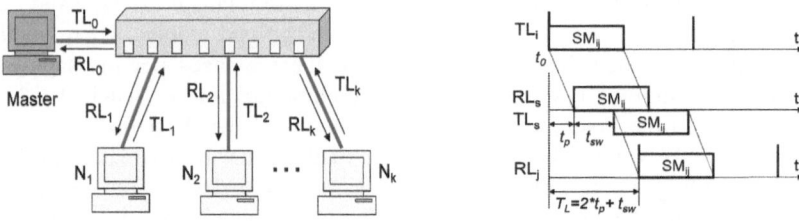

Fig. 1. Maser-slave-based switched Ethernet   **Fig. 2.** Message switching on switched Ethernet

In the cut-through switching, which is used for fast delivery of frames, the switch decides the output port right after receiving the header of a frame and forwards the frame to the output port. When a switch uses the cut-through switching mode, if a message from node i to node j begins to be transmitted at time t_0, the first bit of the message arrives at node j after $T_L = 2*t_p + t_{sw}$ amount of delay from t_0 if there is no collision at the output port and the output queue is empty. This is shown in Fig. 2. Here, t_p is a propagation delay on a link between a node and the switch and t_{sw} is the switching latency (destination port look-up and switch fabric set-up time).

As shown in Fig. 3, in our synchronized switched Ethernet, all of the links consist of a set of consecutive Macro Cycles (MCs), and each MC consists of *Sync* field and a set of consecutive Elementary Cycles (ECs) which is similar to that of FTT-CAN [2]. Each MC has $L = LCM(P_{ij})$ ECs, where P_{ij} denotes the period of a synchronous message SM_{ij} and $LCM(P_{ij})$ denotes the least common multiple of the periods of all of the synchronous messages. *Sync* filed is used to synchronize all of the slave nodes, and includes information about the message transmission model. Message transmission on the switched Ethernet is triggered by a master node which sends a *TM* (Trigger Message) at the beginning of every EC. An EC consists of a *TM*, a *SMP* (Synchronous Message Period) for transmitting periodic messages and an *AMP* (Asynchronous Message Period) for transmitting aperiodic messages. In this paper, we only consider the scheduling of periodic messages. The *TM* contains a message schedule for the periodic messages that are transmitted on the respective EC.

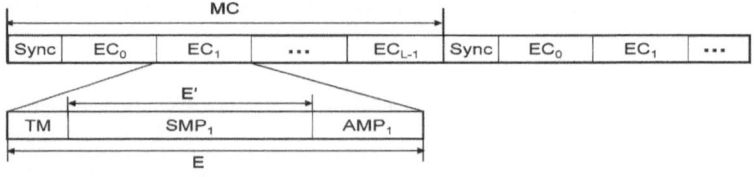

Fig. 3. Message transmission model on the synchronized switched Ethernet

Each periodic (or synchronous) message from node i to node j, SM_{ij}, has a real-time requirement that are characterized by (D_{ij}, P_{ij}, C_{ij}) where D_{ij}, P_{ij}, C_{ij} are the deadline, period, and length of SM_{ij}, respectively. We assume that all the D_{ij} and P_{ij} are the multiple of E and $P_{ij} = D_{ij}$. Thus, the real-time requirement of a periodic message SM_{ij} can be represented by $RT_{ij} = (P_{ij}, C_{ij})$.

3 EDF-Based Real-Time Message Scheduling of Periodic Messages on the Synchronized Switched Ethernet

In this section, a feasible condition for real-time scheduling of periodic messages on the switched Ethernet and an EDF-based real-time message scheduling algorithm which satisfies the condition are described.

When a message is transmitted on a transmission link, it can appear on the reception link after T_L amount of time (Fig. 2). We define the following notations for the description of the feasible condition and scheduling algorithm.

- UL_i and UL_j are the utilization of TL_i and TL_j such that

$$UT_i = \sum_{j \in ST_i} \frac{C_{ij}}{P_{ij}} \qquad UR_j = \sum_{i \in SR_j} \frac{C_{ij}}{P_{ij}} \qquad (1)$$

  where ST_i is a set of nodes to which node i sends its messages and SR_j is a set of nodes from which node j receives the messages.
- $UT_{max,j}$ is the maximum utilization of a set of transmission links that transmit messages to node j, such that

$$UT_{max,j} = \max_{i \in SR_j}\{UT_i\} \qquad (2)$$

- $T_{i,n}$ is the total time of messages which are transmitted on TL_i in n^{th} EC.

3.1 Feasible Condition of Periodic Messages on the Switched Ethernet

The feasible condition for real-time scheduling of periodic messages on the switched Ethernet is in Theorem 1.

Theorem 1. Let C be a set of periodic messages such that $C = \{SM_{ij}\}$ with $RT_{ij} = (P_{ij}, C_{ij})$. C is feasible on the switched Ethernet if it satisfies the following condition for every message SM_{ij} in C,

$$UT_i + UR_j \leq \frac{E' - 2 * \max\{C_{ij}\} + \min\{C_{ij}\}}{E} \qquad (4)$$

Proof. See [9].

An algorithm to check the scheduling condition of Theorem 1 in the master node is described in [9].

3.2 EDF-Based Real-Time Message Scheduling Algorithm on the Switched Ethernet

The master node receives the dynamic changes of the real-time applications from the slaves in AMP and checks the scheduling condition and makes a message transmission schedule for the next MC on the switched Ethernet. The master node transmits a list of messages to be transmitted on each EC of the MC in the *TM* message which is transmitted at the beginning of each EC. When a slave node receives a *TM* message, it interprets the message and transmits the messages specified in the *TM* message. The following *edfScheduling* algorithm is an EDF-based scheduling algorithm which makes a transmission schedule for a set of periodic messages which satisfies the feasibility condition of Theorem 1.

```
Algorithm edfScheduling (schedBuf, MS)
// schedBuf: sorted message buffer to check feasibility
// MS(n): message transmission schedule for n-th EC
// Cmin, Cmax: minimum and maximum length of messages
// r(k): variable to check the readiness of k-th message
//          r(k) = 1 if k-th message is ready
// Tmax(i): the latest finishing time of messages on TLi
// Rmax(j): the available time to send messages on RLj in the worst case
1. for (i = 0; i < N; i++)  {
2.      Tmax(i) = UTi * E + Cmax;
3.      maxTL = max{UTj} for all j where TLj sends messages to RLi;
4.      Rmax(i) = E' − maxTL * E − Cmax + Cmin;
5. }
6. for (k = 0; k < M; k++)   r(k) = 1;
7. for (n = 0; n < L; n++)  {   // L: LCM(Pij)
8.      MS(n) = NULL;
9.      for (i = 0; i < N; i++)  { TLi = RLi = 0; }
10.     for (k = 0; k < M; k++)  {
11.         if (r(k) == 1)  {
12.             SMij = k-th message in schedBuf;
13.             if ((TLi + Cij) ≤ Tmax(i)) && (RLj + Cij) ≤ Rmax(j))) {
14.                 Add SMij to MS(n);
15.                 TLi = TLi + Cij;
16.                 RLj = RLj + Cij;
17.                 r(k) = 0;
18.             }
19.         }
20.         if ((n+1) mod Pij) == 0)  r(k) = 1;
21.     } // end of inner for
22. } // end of outer for
```

In *edfScheduling* algorithm, $T_{max}(i)$ denotes the latest finishing time of messages transmitted on TL_i which denotes the latest time that can be transmitted without violating the boundary of the corresponding EC on RLs. $R_{max}(i)$ denotes the latest finishing time of messages transmitted on RL_i which denotes the latest time within which all of the messages arrived at RL_i have to be transmitted completely (see [9] for more details). The following theorem proves that all of the messages in a feasible message set satisfying the condition of Theorem 1 can be delivered within their deadlines if we schedule those messages according to the proposed *edfScheduling* algorithm.

Theorem 2. The EDF-based scheduling algorithm, *edfScheduling*, guarantees the timely delivery of messages for a message set satisfying the scheduling condition of Theorem 1.

Proof. See [9].

4 Performance Evaluation

This section describes the performance evaluation of the proposed real-time protocol on the switched Ethernet.

Firstly, we have evaluated whether all of the messages that passed the feasibility condition of Theorem 1 are actually transmitted within their deadlines. The simulation parameters are shown in Table 1. The switched Ethernet has 10Mbps data rate and 1 master node and 10 slave nodes on the network. We have generated messages with random source and destinations. Message length was chosen randomly from 100 to 200Bytes. The period of each message was chosen randomly from 1, 2, 3, 4, 6, and 12, so the MC size is 12.

Table 1. Simulation parameters

Bandwidth	10Mbps
Min/Max size of a message	100Bytes / 200Bytes
EC length	1,000us
SP length	900us
MC size	12 ECs

Our message scheduling algorithm is dynamic, which can add new periodic messages while transmitting real-time messages scheduled already. At each MC, the master node collects the message specifications of the messages newly added and checks the feasibility of those messages without affecting the real-time transmission of the messages that had been scheduled already. To show the dynamic feature of the proposed algorithm, we generated 40 messages initially (at MC 0), and checked the feasibility of the messages. All of the initial 40 messages were feasible. After that, we continue to add 10 new messages at each MC while transmitting the messages that were scheduled already until the minimum drop ratio becomes lower than 20%. The simulation result showed that the number of feasible messages increased from 40 (at MC 0), 49 (at MC 1), 55 (at MC 2), 61 (at MC 3), 67 (at MC 4), 71 (at MC 5), 73 (at MC 6), 75 (at MC 7). The number of feasible messages increases as the MC goes by, but the number of newly admitted messages decreases because the link utilization of the switched Ethernet becomes high as more messages are added newly.

To show whether all of the feasible messages are transmitted within their deadline, we have calculated the response time of each message instance of feasible messages. The response time of a message instance is the elapsed time from when the message was generated to the time when the message arrived at the destination. Fig. 4 shows the response time for the messages that were feasible at MC 7. As shown in Fig. 4, all of the message instances were transmitted within their deadline.

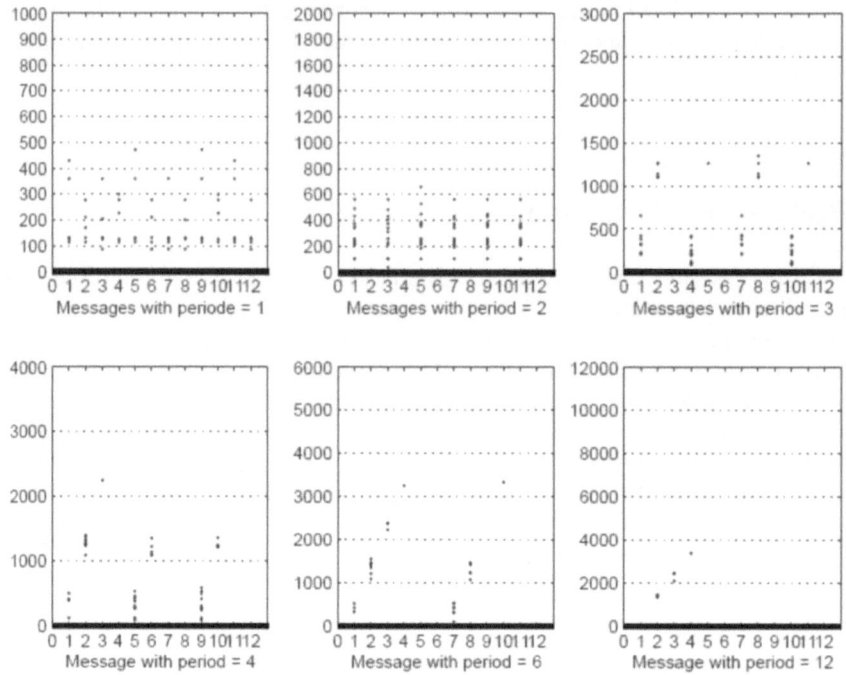

Fig. 4. Response times for the messages that were feasible at MC 7

Secondly, we have compared the real-time capacity between the switched Ethernet and the original Ethernet. The switched Ethernet with N nodes consists of N transmission links and N reception links. Thus, the message transmission capability of the switched Ethernet will be higher than that of Ethernet. However, how much is the communication capability of the switched Ethernet depends on the traffic pattern of the messages in the head of the queue of each node. If all of the messages in the head of the queue are destined to the same node(as shown in Fig. 5-(a)), then the communication capability of the switched Ethernet will almost be the same as that of Ethernet due to the collision at the reception link. On the contrary, if all of the messages in the head of the queue are destined to the different nodes (as shown in Fig. 5-(b)), then all of the messages can be transmitted at the same time, thus the message transmission capability of the switched Ethernet will be N times higher than that of Ethernet.

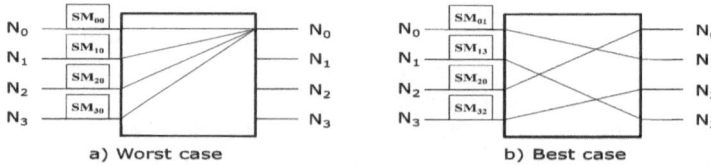

Fig. 5. Message transmission capability of a switched Ethernet depending on the traffic pattern

To compute the message transmission capability of the switched Ethernet, we generated 500 messages randomly, and checked the feasibility condition of Theorem 1 from the first message increasing the number of messages one by one until a certain percentage of message drop occurs and calculated the number of feasible messages. In the case of the Ethernet, we checked the feasibility condition in [7] in the same way. We have done the same kind of simulation 10 times and calculated the averages number of feasible messages. Fig. 6 denotes the message transmission capability of the switched Ethernet and Ethernet. *Single-Ethernet* and *Single-Switch* denotes when we have increased the number of messages until the first message drop occurred in the Ethernet and the switched Ethernet, respectively, and *10%-Ethernet* and *10%-Switch* denotes when we have increased the number of messages until 10% message drop occurred. In the case of the Ethernet, the number of feasible messages has increased just a little when we increased the percentage of message drop, but was almost the same when the number of nodes on the network was increased. On the contrary, in the case of the switched Ethernet, the number of feasible messages has increased as the percentage of message drop was increased. Also the number of feasible messages has increased linearly as the number of nodes on the network was increased.

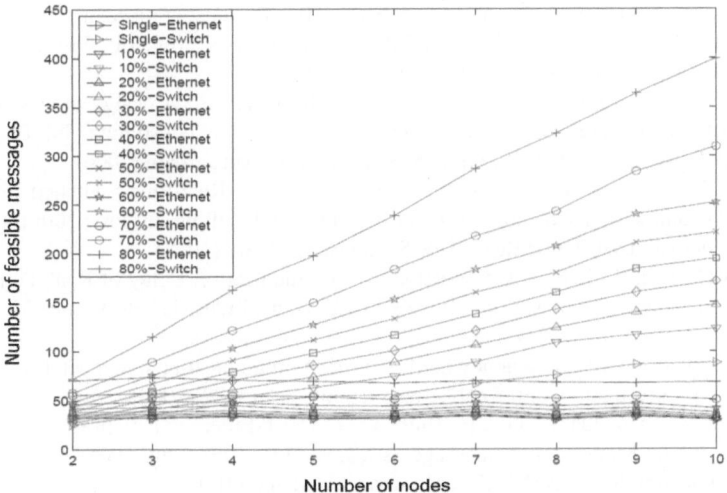

Fig. 6. Comparison of the message transmission capability

5   Conclusions and Future Works

Real-time distributed control systems have been widely used in many industrial applications, such as process control, factory automation, vehicles and so on. In these applications, each task must be executed within a specified deadline, thus, the communications between the tasks have to be completed within their deadlines to satisfy the real-time requirements. Switched Ethernet which is the most widely used in the office has certain good operational features for real-time communications. However, it needs some mechanisms to regulate the traffic on the network in order to satisfy the hard real-time communication requirements of the industrial applications.

In this paper, an EDF-based scheduling algorithm for hard real-time communication over switched Ethernet was proposed. With this scheduling algorithm, there is no need to modify the original principles of switches to support hard real-time communication in industrial environment. This paper also analyzed the scheduling condition for real-time periodic messages and showed that the proposed scheduling algorithm reflects correctly the feasibility condition of the periodic messages on the switched Ethernet. The proposed scheduling algorithm can handle dynamic real-time messages. When some messages need to be added or be deleted, the master node checks the scheduling condition for the updated message set and makes a transmission schedule for the newly updated message set.

Acknowledgments. This research was supported by the 2007 Research Fund of University of Ulsan.

References

1. Mifdaoui, A., Frances, F., Fraboul, C.: Real-time Communication over Switched Ethernet for Military Applications. In: ACM CoNEXT 2005, Toulouse, France, pp. 195–197 (2005)
2. Almeida, L., Pedreiras, P., Fonseca, J.A.: The FTT-CAN protocol: Why and how. IEEE Trans. Industrial Electronics 49, 1189–1201 (2002)
3. Georges, J.P., Krommenacker, N., Divoux, T., Rondeau, E.: A design process of switched Ethernet architectures according to real-time application constraints. In: Eng. Appl. of Artificial Intelligence, vol. 19, pp. 335–344. Elsevier, Amsterdam (2006)
4. Hoang, H., Jonsson, M., Hagstrom, U., Kallerdahl, A.: Real-time Switched Ethernet with earliest deadline first scheduling protocols and traffic handling. In: Proc 10th Int. Workshop on Parallel and Distributed Real-Time Systems, FL, USA (2002)
5. Lee, K.C., Lee, S., Lee, M.H.: Worst Case Communication Delay of Real-Time Industrial Switched Ethernet With Multiple Levels. IEEE Trans. Industrial Electronics 53, 1669–1676 (2006)
6. Loser, J., Hartig, H.: Low-latency hard real-time communication over switched ethernet. In: Proc. 16th Euro-micro Conf. Real-Time Systems, pp. 13–22 (2004)
7. Pedreiras, P., Almeida, L., Gai, P., Buttazzo, G.: FTT-Ethernet: a platform to implement the Elastic Task Model over message streams. In: IEEE Int. Workshop on Factory Communication Systems (WFCS), Vasteras, Sweden (2002)
8. Varadarajan, S., Chiueh, T.: EtheReal: A Host-Transparent Real-Time Fast Ethernet Switch. In: Proc. of Int. Conf. on Network Protocols (ICNP), Austin, TX (1998)
9. Cuong, D.M., Kim, M.K.: Real-time Communications on an Integrated Fieldbus Network Based-on a Switched Ethernet in Industrial Environment. In: Lee, Y.-H., Kim, H.-N., Kim, J., Park, Y.W., Yang, L.T., Kim, S.W. (eds.) ICESS 2007. LNCS, vol. 4523, pp. 357–368. Springer, Heidelberg (2007)

Virtual Container Based Consistent Cluster Checkpoint

Xiao-jia Xiang[*], Hong-liang Yu, and Ji-wu Shu

Department of Computer Science and Technology,
Tsinghua University, Beijing, China
Xiangxj05@mails.tsinghua.edu.cn, {hlyu,shujw}@tsinghua.edu.cn

Abstract. Checkpoint can store and recovery applications when faults happen and is becoming critical to large information systems. Unfortunately, existing checkpoint tools have some limitations such as: not transparent to applications, ignoring file system states, cluster checkpoint is not well supported, and so on. We present a light weight virtual container based cluster checkpoint. Firstly, a virtual container, IPG (Isolated Process Group), is designed to wrap all target applications together and produce checkpoint transparently and completely. Secondly, each IPG has its independent namespace built on an exclusively owned LV (Logical Volume), which can be checkpointed synchronously with the IPG's memory to guarantee the consistency. Finally, distributed applications can be deployed on many IPGs and a cluster checkpoint protocol is presented to orchestrate all IPGs to produce global checkpoints. Experiments and evaluations results illustrate that no overhead will be introduced for applications running in IPGs, and our prototype system works more stable than the library base checkpoint tools.

Keywords: IPG (Isolated Process Group), LV (Logical Volume).

1 Introduction

The need for fault tolerance in today's large scale information systems is becoming critical. Many computing applications, especially those in cluster environment, which may run several days or weeks, can't tolerate system fault such as power outage, data corruption, and so on. Unfortunately, due to the growing complexity of today's computing systems and the extraordinarily large number of faults ranging from transient errors to malicious failures [1], the MTBF of computing systems may be shorter than the execution times of the running scientific computing applications.

Checkpoint is a suitable solution for this problem by saving applications' running states periodically on stable storage pool. Restoring applications can load the saving states and continue to run from just the latest checkpoint time. Most of studies have been conducted on checkpoint, however, there existing some limitations.

First of all, many checkpoint tools are not transparent to applications. ftIO [2] implements a file operation wrapper layer with which a copy-on-write file replica will

---

[*] Supported by National Grand Fundamental Research 973 Program of China (Grant No. 2007CB311100).

D. Ślęzak et al. (Eds.): CA 2009, CCIS 65, pp. 71–78, 2009.

be generated while keeping old data unmodified. MOB [3] and Metamori [4] also wrap standard file IO operations to buffer file changes between checkpoints. In order to user these wrapper layer, applications have to be modified. Zap [5] present pod, that is processes domain, to decouple the protected processes from dependencies to the host operating system. A thin virtualization layer is inserted above the OS to support checkpoint while no application modification is needed. We borrow the idea of Zap to construct our own process container.

Secondly, most of checkpoint tools focus on taking memory image of one or several processes, they are helpless facing the scenario where many processes communicate frequently and should dump their states altogether during checkpoint. Libckpt [6] is an open source portable checkpoint tool for Unix, it mainly focus on performance optimization, while only supports single-threaded processes checkpoint. Libtckpt [7] and BLCR [8] improve the mechanism to spawn a separate thread to support multiple threads checkpoint, however, no solution to checkpoint related processes was exploited.

Thirdly, in most of the checkpoint tools, no mechanism is produced to guarantee the consistency between the applications' memory images, which are the main part dumped in checkpoint including all heaps, stacks, sockets, pipes, etc, and their file system states, which are always be neglected. Condor [9] assumes that each file is only opened in append mode, records the file length upon checkpoint, truncates it on recovery. This method is not good enough for other file changes, such as data update, file attributes modification. ReFS [10] introduces an address translation layer into the kernel to extend ext2 file systems with multi-version supporting, which can save old data when file is changed. ReviveI/O [11] uses hardware to buffer data, and is implemented as a pseudo device driver. Zap [5] deals with the memory and file system data consistency problem by utilizing the network storage hardware, such as NAS server or SAN storage. The assumption is there is no any fault happening on the network storage.

Finally, cluster checkpoint is seldom covered due to the difficulties to dump network states and keep communication peers alive after restoration. Cruz [12] builds on Zap and implements a coordinate protocol to ensure global consistent checkpoint. In order to dump and save network states, many network related system calls are added or changed to monitor the socket send/receive buffers. ZapC [13] also build on Zap, it designs a two stage coordinate protocol for cluster checkpoint and the first network state checkpoint stage is time consuming. Furthermore, a manager node must be explicitly introduced to orchestrate the global checkpoint among all pods which are all become the slaves of the manager during checkpoint by integrating an agent module. However, the overhead of adding a new manager node can't be neglect, and the reliability of this mechanism is not discussed.

In this paper, we present a virtual container based consistent cluster checkpoint. Firstly, the virtual container, we called IPG (Isolated Process Group), is actually a tightly coupled group of processes with isolated storage name space and independent network interface. It is a thin virtual running context which constructs an ideal environment for any application executing within it. Furthermore, no matter how many processes running, an IPG can freeze and dump all of them during checkpoint transparently and consistently. Secondly, each IPG has its own name space which resides on a logic volume pre-allocated from the physical storage pool. Consistent checkpoint achieves by not only saving the memory images, but also saving the IPG's logic volume simultaneously. Different from pod in Zap, IPG don't rely on the network storage, and no assumptions of reliability are hold. Finally, cluster applications can be deployed in many IPGs which embed cluster module transparently and consist of a reliable quorum automatically. A cluster checkpoint protocol is presented to orchestrate all IPGs to produce a global consistent checkpoint efficiently without the need of additional manager node.

The paper is organized as follows. System overview is discussed in section 2. The detail of design and implementation is presented in section 3. Section 4 provides the evaluation of our prototype system, then we conclude in section 5.

2 System Overview

The system architecture is illustrated in figured 1. Left side is the unprotected physical node wherein three applications, that is, oracle, video-lan vod server, and mysql, are running without any fault tolerance mechanism. After introducing a virtualization layer, each application can be executed separately in its own container with checkpoint supported as shown in the right side of figure 1. Therefore, IPG checkpoint mechanism can be used to checkpoint all application processes. This translation is transparent to applications. In addition, if these three applications construct a distributed service, their containers, that is, the IPGs, can be configured as a high reliable cluster quorum and can cooperate to get global consistent checkpoints.

As shown in figure 1, our system consists of three main parts: OS virtualization layer, IPGs, the checkpoint daemon and utilities. The OS virtualization is the low level foundation for IPG. It is mainly responsible for IPGs scheduling, supervising and tracking IPGs' IO and network stream, and most importantly, providing IPG freezing and dumping mechanism to support checkpoint. The IPGs is the thin virtual

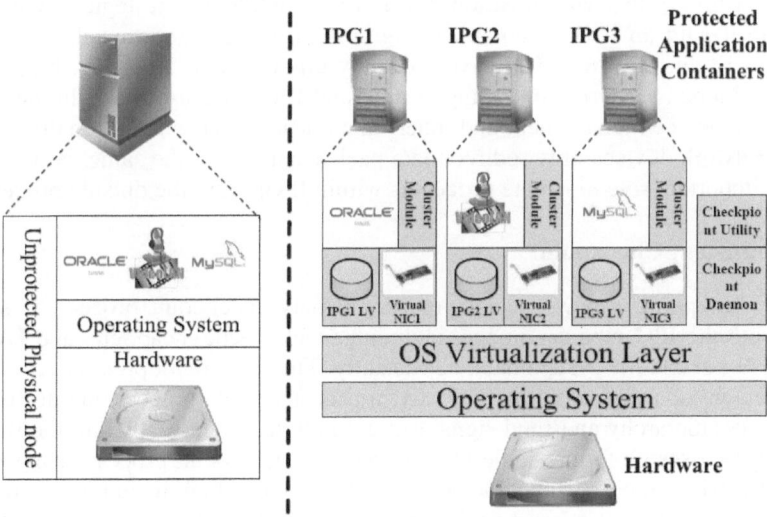

Fig. 1. System architecture

executing context for applications. Each IPG is outfitted with three module/resources: IPG's cluster module communicates with other peers, tracks IPG's state in cluster, cooperates to fulfill cluster checkpoint according to protocol. IPG LV, that is, logical volume, is the virtual storage resource exclusively owned by the corresponding IPG, whose physical constitution is determined by virtualization layer and is transparent to inner applications. A virtual NIC is a pseudo device layer which monitors the IPG's input/output network data packets and redirects them to the right destination. Checkpoint daemon communicates with the virtualization layer to trigger checkpoint periodically and adjust the interval automatically. Checkpoint utility is a user level toolkit which provides an interface for users to control the checkpoint, manage the dumped memory images and IPG's LV replicas.

3   Design and Implementation

3.1   IPG: Light Weight Process Level Application Container

In our system, the protected application container, i.e. IPG, is light weight virtualized execution environment with checkpoint supported. It based on the thin virtualization layer to provide a standard view of operating system to the applications within it. The IPG is light weight because all application processes running in it can access system resources directly just like those outside the container, little overhead will be introduced.

There are several implementation issues should be discussed. Firstly, IPG is actually made up of a bundle of processes which are originally forked from the external processes. As an operating system, each IPG owns its 'init' process and inherits the process hierarchy structure. Secondly, processes within IPGs share the same memory management mechanism with the external processes because these IPG processes are all visible from outside, owning external PID, and the IPG process hierarchy is actually part of the global hierarchy of the physical node's operating system. Thirdly, the virtualization layer will pre-allocate a logical volume and associate it with an IPG. A complete file system will be made on this LV, and all processes in this IPG will exclusively occupy this resource. Fourthly, block devices are shared among external operating system and IPGs because the light weight IPGs provide no device management and inherit the outside device table. Finally, the IPG's virtual network devices only redirect data packets and share the same hardware. They play an important role to isolate processes within IPGs from the outside processes.

3.2   Consistent Checkpoint

During checkpoint, a new special process, namely checkpoint process, is spawn to dump both the IPG memory image, including stack, cache, heap, etc, and LV, which holds IPG's stable file system data, consistently. The checkpoint process can change its executing environment arbitrarily. Firstly, it acts as an IPG process, iterates the whole IPG process hierarchy and send signal to freeze all the processes within the IPG. After all IPG processes go to the stable frozen state, the checkpoint process changes its role to be an outside process which and access all the storage pool, including the IPG's LV. Finally, the checkpoint process replicates the frozen IPG LV from the block level and builds mapping relationship between the dumped memory image and the LV replica.

In order to use the storage resource efficiently, de-duplication will be done among checkpoints taken at different time.

3.3 Supporting for Cluster

In order to support cluster applications, each IPG is outfitted with a cluster module to communicate and act as a physical node. The cluster module will spawn a process to probe other IPGs' states and determine what role the owner IPG should play in the cluster. This process is triggered every time the IPG is created and should not be checkpointed.

With the help of cluster module, all IPGs will construct a quorum to cooperate for cluster checkpoint. Each IPG will play one of the three following roles in the quorum: master, slave, and arbitrator. The master is the maintainer of the quorum and is responsible for sending heartbeats, managing all quorum members, initiating and orchestrating the cluster checkpoint, and so on. The slave is only a participator of the quorum. It mainly responses the master's requests, initiating the its owner IPG's checkpoint. The arbitrator is an unstable role, which only appears in master choosing procedure and is actually the master candidate with not enough votes. This quorum is high reliable, the global consistent checkpoint will be available as long as the majority of the IPGs are alive and can communicate normally.

Our cluster checkpoint protocol is based on TPC (Two Phase Commit) protocol, the master is the coordinator and the whole procedure is described in figure 2.

Master	Slave
1 Send 'Checkpoint' to all slaves 2 Shut down Virtual NIC output channel, send 'freeze' signal to all processes of its owner IPG 2a Prepare memory image dump file and IPG LV replica 3 Wait to Receive 'Prepared' reply from all slaves 4 If receive all 'Prepared' before timeout, send 'goon' to all slaves; else send 'abort' to all slaves 5 If checkpoint will go on, do the following dumping and replicating, else resume the IPG, end the procedure 6 After checkpoint IPG succesfully, wait for slave status 7 receive all 'done' from slaves, send 'continue' to slaves, and resume the owner IPG	1 Receive 'Checkpoint' from master 2 Shut down Virtual NIC output channel, send 'freeze' signal to all processes of its owner IPG 2a Prepare memory image dump file and IPG LV replica 3 Send 'Prepared' to the master and wait for reply 4 If receive 'goon', then do the following dumping and replicating, else resume the IPG, end the procedure 5 After checkpoint IPG succesfully, send 'done' to master 6 Wait for master until the 'continue' is received, then resume the owner IPG

Fig. 2. Cluster Checkpoint Procedure

4 Evaluation

In this section, we first evaluate the overhead introduced by our light weight IPG. Then, we compare the checkpoint performance between our prototype system and the open source library-based checkpoint tools ckpt [14].

The physical node used in our test is outfitted with an Intel Core(TM) 2 CPU T7200@2.00GHz, 2GB RAM, and 160GB Seagate SATA disk drives. The OS of the node is Linux with 2.6.18 kernel.

4.1   The Overhead of IPG

In this experiment, We use bonnie++ to do five micro tests step by step outside and inside IPG separately: (1)ByteW: writing a file with putc() stdio macro. Data is written with the unit of a character. (2)BlockW: writing a file with write() in libc. Data is written with the unit of a block. (3)ReWrite: data in files is read using read() first, then dirtied, and rewritten with write(). (4)ByteR: reading a file with getc() stdio macro. (5)BlockR: reading a file with read(), this should be a very pure test of sequential input performance.

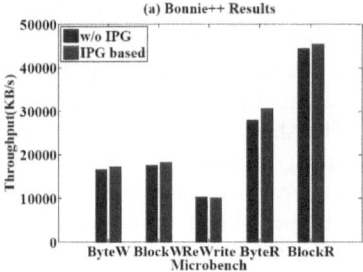
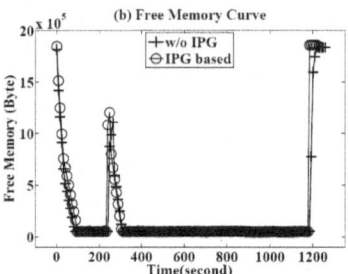

Fig. 3. Overhead Experiments Results

Figure 3(a) shows the experimental results. Except the rewriting, benchmarks running in the IPG outperform those running outside. As the results shows, for ByteW, throughput inside IPG increase 3.33% comparing with that outside IPG; for the other benchmarks, the increasing ratios are 3.54%, -1.48%, 9.65%, 2.55% separately. The reason for even better performance in IPG partly dues to the different computing/IO resources scheduling policy introduced by the OS virtualization layer.

We also monitor the system performance from the resource view. We record the CPU and memory usage during the whole experiment. For both tests inside and outside IPG, the CPU usage ratios are always 100% due to computing intensive characteristic of Bonnie++. Figure 3(b) depicts the memory usage statistics. We can observe that fluctuation of both curves is similar and the whole procedure inside IPG ends before the outside one by 39 seconds. If we use the average free memory as a criteria to measure the memory consuming status, we can get that the average free memory is about 195237.15 bytes with IPG, while that for w/o IPG is 208579.78 bytes. The conclusion is that introducing IPG only consumes more 13Kbytes memory in this experiment.

From aforementioned data, we can conclude that little overhead will be introduced for applications running in IPG. The reason is that applications in IPG can also access all resources directly without interrupting by the light weight virtual container.

4.2   Performance Comparison

In this experiment, we compare the checkpoint time between our prototype system and ckpt. We use several applications as benchmarks. After these applications run a constant period of time and become stable, we trigger the checkpoint and record the time cost on checkpointing. These applications are: (1) Sar, a system state monitor; (2) Mem-bench, a tool shipped be the ckpt, which only applies for 100KBytes memory every second, the longer it runs, the bigger the memory it occupies. (3) Kernel building, that is, making a linux 2.6 kernel. (4) Mysql, we runs the mysqld daemon to provide the database service. (5) Videolan, we runs the videolan server and single-cast a video stream to a student PC in our laboratory using UDP protocol.

Table 1. Checkpoint Time Comparison

benchmarks	Ckpt (second)	IPG based (second)
sar	0.073297	0.169008
Mem-bench@50sec	0.092274	0.239921
Mem-bench@100sec	0.120985	0.243374
Mem-bench@150sec	0.156433	0.247091
Kernel building	45.570797	0.441495
Mysql	-	0.62492
VideoLan	-	0.337105

Table 1 illustrate the experiment results. The checkpoint time fluctuate largely for ckpt. For sar and mem-bench running 50, 100, and 150 seconds, comparing with our system, the time cost by ckpt is only 43.37%, 38.46%, 49.71%, 63.31% separately. However, the time cost by ckpt is even 2 order of magnitude longer than us when checkpointing the kernel building application. This is because the library based checkpoint tool ckpt is not good enough to tackle with multiple processes applications. For the IO intensive database, i.e., mysql, and the network traffic intensive video server, i.e., videolan, the ckpt can't even work successfully because there may exist some conflicts between these applications and the library hijack mechanism of ckpt. On the other hand, our system do well in checkpointing mysql and videolan and all applications list in table 1 can be checkpointed within 1 second. In short, our virtual container based checkpoint is more stable and the performance is acceptable.

5   Conclusion

In this paper, we present a light weight virtual container based cluster checkpoint. By designing the light weight container IPG, all application processes can be wrapped and checkpointed completely, in addition, all these happen transparently to applications. Each IPG has its independent namespace built on an exclusively owned LV. By saving not only the memory images, but also the IPG's LV, checkpoint consistency can be guaranteed. A cluster checkpoint protocol is also presented in this paper to support taking global checkpoint from all IPGs where distributed applications are deployed. Experiments and evaluations results illustrate that no overhead will be introduced for applications running in IPGs, our prototype system is stable and the checkpoint performance is acceptable.

References

1. Choy, M., Leong, H.V., Wong, M.H.: Disaster recovery techniques for database systems. Communications of the ACM 43(11) (2000)
2. Lyubashevskiy, I., Strumpen, V.: Fault-tolerant file-I/O for portable checkpointing systems. The Journal of Supercomputing 16(1-2), 69–92 (2000)
3. Pei, D.: Modification Operations Buffering: A Lowoverhead Approach to Checkpoint User Files. In: IEEE 29th Symposium on Fault-Tolerant Computing, Madison, USA, June 1999, pp. 36–38 (1999)
4. Jeyakumar, A.R.: Metamori: A library for Incremental File Checkpointing. Master?s thesis, Virgina Tech, Blacksburg, June 21 (2004)
5. Osman, S., Subhraveti, D., Su, G., Nieh, J.: The Design and Implementation of Zap: A System for Migrating Computing Environments. In: Proceedings of the Fifth USENIX Symposium Operating Systems Design and Implementation, Boston, MA, USA, December, 2002, pp. 361–376 (2002)
6. Plank, J.S., Beck, M., Kingsley, G., Li, K.: Libckpt: Transparent Checkpointing Under Unix. In: Proceedings of the USENIX Winter 1995 Technical Conference, New Orlands, LA, USA, January 1995, pp. 213–223 (1995)
7. Dieter, W.R., Lumpp Jr., J.E.: User Level Checkpointing for Linux Threads Progams. In: Proceedings of the FREENIX Track: 2001 USENIX Annual Technical Conference, Boston, MA, USA, June 2001, pp. 81–92 (2001)
8. Duell, J., Hargrove, P., Roman, E.: The Design and Implementation of Berkeley Lab's Linux Checkpoint/Restart. White paper. Future Technologies Group (2003)
9. Litzkow, M., Tannenbaum, T., Basney, J., Livny, M.: Checkpoint and Migration of UNIX Processes in the Condor Distributed Processing System. Technical Report CS-TR-1997-1346, University of Wisconsin, Madison (April 1997)
10. Kim, H., Yeom, H.: A User-Transparent Recoverable File System for Distributed Computing Environment. In: Proceedings of CLADE 2005, July 2005, pp. 45–53 (2005)
11. Nakano, J., Montesinos, P., Gharachorloo, K., Torrellas, J.: RevivoI/O: Efficient Handling of I/O in Highly-Available Rollback-Recovery Servers. In: Proceedings of HPCA 2006, Austin, Texas, USA, February 2006, pp. 200–211 (2006)
12. Janakiraman, G., Santos, J.R., Subhraveti, D., Turner, Y.: Cruz: Application Transparent Distributed Checkpoint Restart on Standard Operating Systems. In: Proceedings of DSN 2005, Yokohama, Japan, 28 June-1 July, 2005, pp. 260–269 (2005)
13. Laadan, O., Phung, D., Nieh, J.: Transparent Checkpoint Restart of Distributed Applications on Commodity Clusters. In: Proceedings of the 2005 IEEE International Conference on Cluster Computing, Boston, MA, USA, September 2005, pp. 1–13 (2005)
14. Zandy, C.: Ckpt – Process Checkpoint Library, http://pages.cs.wisc.edu/~zandy/ckpt/

Terminal Sliding Mode Control of DC-DC Buck Converter

Chian-Song Chiu[1,*], Ya-Ting Lee[2], and Chih-Wei Yang[1]

[1] Department of Electrical Engineering, Chung-Yuan Christian University,
Chung-Li 32023, Taiwan
`cschiu@dec.ee.cycu.edu.tw`
[2] Department of Beauty Science, Chienkuo Technology University,
Changhua 500, Taiwan
`ytlee@ctu.edu.tw`

Abstract. This paper presents a novel sliding mode control for DC-DC buck converter. Our study is motivated by the output regulation problem of DC-DC converters with varying load and inaccuracy passive elements. In order to achieve stable output voltage, a terminal sliding mode controller (TSMC) is proposed. Different to traditional sliding mode control, a terminal sliding surface is introduced to assure finite convergent time of control errors on the surface. In addition, due to unknown bound of uncertainty, an adaptive law is derived to compensate the effect. As a result, the TSMC is modified to adaptive TSMC (ATSMC). Furthermore, this controller is implemented by a DSP card. Form the simulation and experiment results, the proposed control scheme is shown to be feasible.

Keywords: DC-DC converter, uncertainty, terminal sliding mode control.

1 Introduction

The main control method in the switching power converter is associated with a pulse width modulation (PWM) [1, 2, 3]. The controlled PWM signal drives the transistor to maintain a constant output voltage, where the duty cycle of the PWM signal is adjusted by the controller. To this end, many control methods have been used, such as PID [4], fuzzy control [5], H_∞ control [6], feedback linearization [7], etc.

Since the sliding mode control has good robustness to system parameter uncertainty and external interference [8], the sliding mode control is widely used in various fields. Many application examples are discussed in literature [9, 10]. For example, in electric fields [11], the sliding mode control is used to inverter, dc motor, induction motor, permanent magnet synchronous motor, etc. The sliding mode control is also applied to power converters [12, 13]. In the classical sliding mode control method usually chooses a linear sliding function to obtain an exponentially stable manifold. In contrast, the work [14] proposed a special terminal sliding mode. Different from the classical sliding mode control, the terminal sliding mode control (TSMC) has a nonlinear sliding function. The

---

* Correspondence addressee.

D. Ślęzak et al. (Eds.): CA 2009, CCIS 65, pp. 79–86, 2009.
© Springer-Verlag Berlin Heidelberg 2009

nonlinear sliding function provides a terminal convergence (finite-time convergence). Due to the better robustness and convergence, some literatures use the terminal sliding mode control for complex systems [15, 16]. However, no result for controlling uncertain buck converter is proposed via terminal sliding mode control.

In this paper, we present a new TSMC method for the DC-DC buck converter. First, the terminal sliding mode controller is proposed by introducing a new terminal sliding function. Since the actual choice of electronic components has some uncertain errors, an adaptive terminal sliding mode controller (ATSMC) is also introduced. These approaches have many attractive properties, including fast finite time convergence and robustness to parameter uncertainties, etc.

2   Model Formulation

In Fig. 1, an equivalent circuit of a buck converter is illustrated. To obtain a precise model, there are some effects needed to be considered below:

(i) The parasitic resistances of the capacitor C and the inductor L are denoted by R_c and R_L, respectively.
(iii) The forward voltage of the power diode is denoted by V_D.
(iii) The static drain to source resistance of the power MOSFET is denoted by R_M.

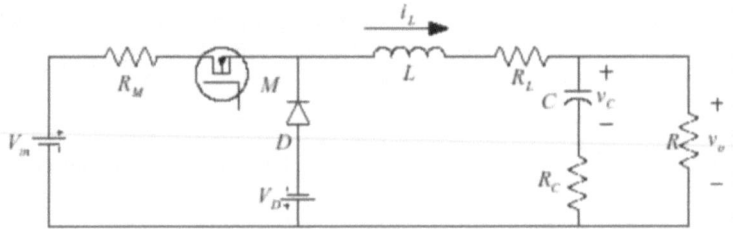

Fig. 1. Equivalent circuit of buck converter

Applying the AM-OTS-DS methodology, the dynamic equations of a buck converter are derived as follows:

$$\dot{X}(t) = f(X) + g(X)d, \tag{1}$$
$$y(t) = CX, \tag{2}$$

where $X = \begin{bmatrix} i_L & v_c \end{bmatrix}^T$; $y(t) = v_o$; $i_L(t)$ and $v_c(t)$ denote the inductor current and capacitor voltage, respectively; $v_o(t)$ is the output voltage; d is the duty ratio of the PWM signal of the buck converter; and other functions are defined below:

$$f(X) = \begin{bmatrix} -\frac{1}{L}[R_L + \frac{RR_c}{R+R_c}] & -\frac{R}{L(R+R_c)} \\ \frac{R}{C(R+R_c)} & -\frac{1}{C(R+R_c)} \end{bmatrix} \begin{bmatrix} x_1 \\ x_2 \end{bmatrix} + \begin{bmatrix} \frac{V_D}{L} \\ 0 \end{bmatrix}$$

$$g(X) = \begin{bmatrix} -\dfrac{1}{L}R_M & 0 \\ 0 & 0 \end{bmatrix} \begin{bmatrix} x_1 \\ x_2 \end{bmatrix} + \begin{bmatrix} \dfrac{1}{L}(V_D+V_{in}) \\ 0 \end{bmatrix}$$

$$C = \begin{bmatrix} \dfrac{RR_c}{R+R_c} & \dfrac{R}{R+R_c} \end{bmatrix}$$

3 Controller Design

A. Terminal sliding mode control

Let the desired voltage value be y_d and define the tracking error as $e(t) = y(t) - y_d$. The control objective is to ensure the tracking error converge to zero as fast as possible. The first step is to define the terminal sliding function as follows:

$$s(t) = e_1 + \frac{1}{\beta}e, \tag{3}$$

where $\dot{e}_1 = e^{q/p}$ (i.e. $e_1 = \int_0^t e^{q/p}(\tau)d\tau$); $\beta > 0$ is a constant; p and q are odd integers satisfying $p > q > 0$. When $s=0$, from Equation (3) we can obtain

$$\begin{cases} \dot{e} = -\beta e^{q/p} & (4) \\ \dot{e}_1 = -\beta^{q/p} e_1^{q/p}. & (5) \end{cases}$$

In order words, the error and the integrate of the error are convergent in finite time. The convergent time of $e(t)$ and $e_1(t)$ are respectively:

$$t_e = \frac{|e(0)|^{1-\frac{q}{p}}}{\beta(1-\frac{q}{p})} \tag{6a}$$

$$t_{e1} = \frac{|e_1(0)|^{1-\frac{q}{p}}}{\beta^{q/p}(1-\frac{q}{p})}. \tag{6b}$$

If we drive the system to the sliding surface $s(t)=0$, the error will converge to zero in finite time. Therefore, we call $s(t)=0$ the terminal sliding surface. The following theorem is given for the TSMC design in the exact model case.

Theorem 1. For the buck converter (1) and (2), if the control input d with the terminal sliding function (3) is designed as:

$$d = \frac{1}{Cg(X)}[-\beta e^{q/p} - \beta K sign(s) - Cf(X)]. \tag{7}$$

where $K, \beta > 0$, $p > q > 0$, then the terminal sliding mode control guarantees finite-time convergent stability.

Proof. For the terminal sliding function (3), its time derivative along the dynamic Equations (1) and (2) is

$$\dot{s}(t) = e^{q/p} + \frac{1}{\beta}\dot{e}$$

$$= (CX - y_d)^{q/p} + \frac{1}{\beta}(Cg(X)d + Cf(X))$$

$$= -K sign(s)$$

where the control law (7) has been applied. Obviously, the condition $s\dot{s} < 0$ is satisfied. Thus, the sliding function will converge to zero in finite time, i.e. $s \rightarrow 0$ for $t \in [0,T]$. Moreover, due to $s=0$, the error $e(t)$ is terminally convergent. Therefore, the error converges to zero in finite time.   ■

B. Adaptive terminal sliding mode control

Since there are parameter uncertainties errors are in the dynamic Equations (1) and (2), the TSMC is not enough. To solve this problem, we propose an adaptive terminal sliding mode controller (ATSMC) to adaptively tune the controller, so that the tracking error be able to achieve zero quickly. To simplify the problem, the terms $f(X)$, $g(X)$ and C are assumed to be uncertain but with known the nominal values. We partition the dynamic functions as

$$Cf(X) = f_c = \hat{f}_c + \Delta f_c \tag{8}$$
$$Cg(X) = g_c = \hat{g}_c + \Delta g_c \tag{9}$$

where Δf_c and Δg_c are assumed to be uncertainties; $\hat{f}_c$ and $\hat{g}_c$ are the nominal parts of f_c and g_c. They satisfy

$$\left| f_c - \hat{f}_c \right| \leq F \tag{10}$$

$$\left| \frac{g_c}{\hat{g}_c} \right| \leq \alpha, \hat{g}_c \neq 0 \tag{11}$$

where the uncertain error of f_c is assumed to be bounded by an upper functions F; and $\alpha > 0$. The ATSMC design is given in the following theorem.

Theorem 2. For the uncertain back converter (1) and (2), the control input d with the terminal sliding function (3) is designed as:

$$d = \frac{1}{\hat{g}_c}[-\beta e^{q/p} - \beta \hat{K} sign(s) - \hat{f}_c] \tag{12}$$

and the adaptive control gain $\hat{K}$ is given by:

$$\dot{\hat{K}} = \frac{1}{\alpha}\xi|s| \tag{13}$$

where $K, \beta > 0$, $p > q > 0$; and $\xi > 0$ is an update gain. Then, the robust output voltage control objective is achieved.

Proof. Select the Lyapunov function candidate as follows: $V = \dfrac{1}{2}s^2 + \dfrac{1}{2}\xi^{-1}\tilde{K}^2$, where $\tilde{K}$ is an adaptation error defined as $\tilde{K} = \hat{K} - K^*$ for the estimate control gain $\hat{K}$ and the desired value K^* satisfying:

$$K^* > \left| \frac{\alpha F}{\beta} + (\alpha-1)\left(e^{q/p} + \frac{1}{\beta}\hat{f}c \right) + \eta \right| \tag{14}$$

with a positive constant η.

Using Equations (12) and (13), we take the time derivative of V as follows:

$$\dot{V} = s\dot{s} + \xi^{-1}\tilde{K}\dot{\tilde{K}}$$

$$= s(e^{q/p} + \frac{1}{\beta}f_c - g_c\hat{g}_c^{-1}e^{q/p} - g_c\hat{g}_c^{-1}\tilde{K}sign(s)$$

$$- g_c\hat{g}_c^{-1}K^*sign(s) - \frac{1}{\beta}g_c\hat{g}_c^{-1}\hat{f}_c) + \xi^{-1}\tilde{K}\dot{\hat{K}}$$

$$= -g_c\hat{g}_c^{-1}|s| + \omega s + \xi^{-1}\tilde{K}\dot{\hat{K}}$$

$$\leq -\eta|s| \leq 0$$

where the condition (14) has been applied such that

$$\omega s = \left| e^{q/p} + \frac{1}{\beta}f_c - g_c\hat{g}_c^{-1}e^{q/p} - \frac{1}{\beta}g_c\hat{g}_c^{-1}\hat{f}c \right| |s| - g_c\hat{g}_c^{-1}K^*|s|$$

$$\leq -\eta|s|$$

Therefore, the convergence of s and $\tilde{K}$ is proven according to Lyapunov stability criterion. We get $\dot{s} \in L_\infty$, $\tilde{K} \in L_\infty$ and $s \in L_2$. By Barbalat's lemma, we obtain $s \to 0$ from $s, \dot{s} \in L_\infty$ and $s \in L_2$. As a result, when $s=0$, it leads to finite time convergence of $e(t)$. Therefore, the control objective is achieved.   ■

4 Simulation and Experimental Results

The buck converter specifications considered in our simulation and experiment are: $V_{in} = 20\text{v}$ and $V_o = y_d = 5$ v. The parameter values are: $L=98.58\text{uH}$, $R_L=48.5\text{m}\Omega$, $C=202.5\text{uF}$, $R_C=162\text{m}\Omega$, $V_D=0.82\text{v}$, $R_M=0.27\Omega$, and $R_{load}=6\Omega$.

When uncertainty is considered, the load resistance R is changed from 6Ω to 16Ω and back 6Ω, and the passive components exist about 10% deviation of nominal values. In this case, Theorem 2 is applied and leads to control results as shown in Fig. 2, while

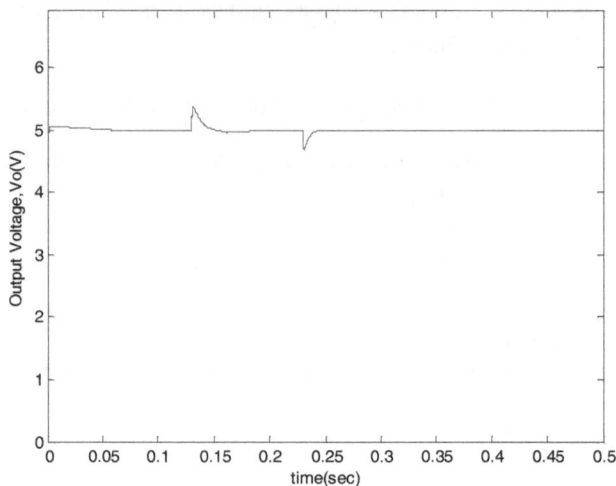

Fig. 2. Output voltage of the buck converter when the load changes from 6Ω to 16Ω and then back 6Ω

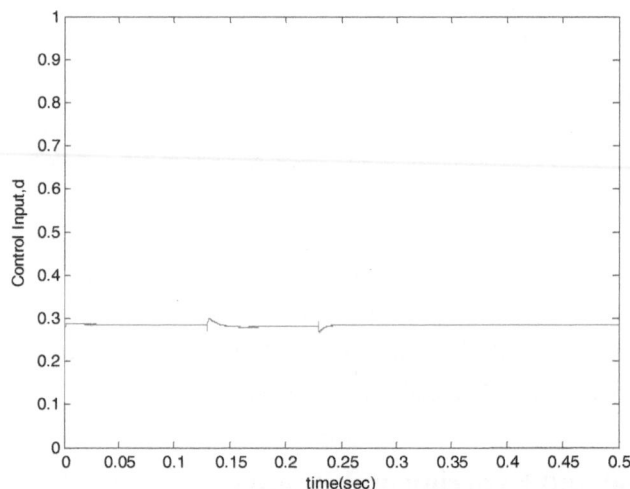

Fig. 3. Control input of the buck converter when the load changes from 6Ω to 16Ω and then back 6Ω

Fig. 3 shows the control input. Furthermore, we apply the ATSMC to the uncertain buck converter by the experiment. The controller is implemented by a DSP card (dSPACE1104). The control responses are shown in Fig. 4. These results demonstrate the fast responses and robustness against parameter variations.

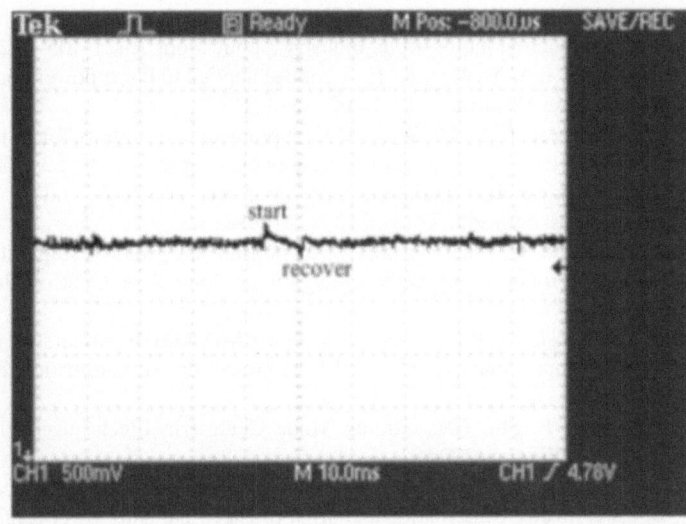

Fig. 4. Experimental output voltage of buck converter when the load changes from 6Ω to 16Ω and then back 6Ω

5 Conclusions

In this paper, we have introduced a terminal sliding mode control to cope with the robust output voltage control problem of a buck converter. After choosing a new integral terminal sliding function, the control error and the integral error converge to zero in finite time on the sliding surface. The terminal sliding manifold guarantees high robustness and fast response. If considering parameter uncertainty, the proposed ATSMC also provides good control performance. The simulation and experimental results have shown the expected performance.

Acknowledgments. This work was supported by the National Science Council, R.O.C., under Grant NSC-97-2221-E-033-059 and Ministry of Economic Affairs, DoIT, R.O.C., under Grant 98-EC-17-A-07-S2-0029.

References

1. EPARC.: Power Electronics. Chinese (2007)
2. May, P., Ehrlich, H.C., Steinke, T.: ZIB Structure Prediction Pipeline: Composing a Complex Biological Workflow through Web Services. In: Nagel, W.E., Walter, W.V., Lehner, W. (eds.) Euro-Par 2006. LNCS, vol. 4128, pp. 1148–1158. Springer, Heidelberg (2006)
3. Foster, I., Kesselman, C.: The Grid: Blueprint for a New Computing Infrastructure. Morgan Kaufmann, San Francisco (1999)
4. He, M., Xu, J.: Nonlinear PID in Digital Controlled Buck Converters. In: IEEE APEC, pp. 1461–1465 (2007)

5. Singh, B.N., Bhim Singh, B.P.: Fuzzy control of integrated current-controlled converter-inverter-fed cage induction motor drive. Industry Applications 35(2), 405–412 (1999)
6. Nairn, R., Weiss, G., Ben-Yaakov, S.: H $_\infty$ control applied to boost power converter. IEEE Transactions on Power Electronics 12(4), 677–683 (1997)
7. Sanders, S.R., Varghese, G.C., Cameron, D.F.: Nonlinear control laws for switching power converters. In: 25th IEEE Conference on Decision and Control, pp. 46–53 (1986)
8. DeCarlo, R.A., Zak, S.H., Matthews, G.P.: Variable structure control of nonlinear multivariable systems: a tutorial 76(3), 212–232 (1988)
9. Slotine, J.J., Sastry, S.S.: Tracking control of non-linear systems using sliding surfaces with application to robot manipulator. International Journal of Control 38(2), 465–492 (1983)
10. Zhihong., M., Paplinski., A.P., Wu, H.R.: A robust MIMO terminal sliding mode control scheme for rigid robotic manipulators. IEEE Transactions on Automatic Control 39(12), 2465–2469 (1994)
11. Utkin, A., Guldner, J., Shi, J.X.: Sliding Mode Control in Electromechanical Systems. Taylor & Francis, Abington (2002)
12. Mayakkannan, A.V., Rajapandian, S.: Modeling and Analysis of Sliding Mode Controller for Buck-Boost Converter. International Journal of Electrical and Power Engineering 2(3), 147–153 (2008)
13. Tan, S.-C., Lai, Y.M., Tse, C.K., Cheung, M.K.H.: An adaptive sliding mode controller for buck converter on continuous conduction mode. In: IEEE APEC, vol. 3, pp. 1395–1400 (2004)
14. Venkataraman, S.T., Gulati, S.: Terminal sliding modes: a new approach to nonlinear control synthesis. In: Fifth International Conference on Advanced Robotics, vol. 1, pp. 443–448 (1991)
15. Feng, Y., Yu, X., Man, Z.: Adaptive fast terminal sliding mode tracking control of robotic manipulator. In: 40th IEEE Conference on Decision and Control, vol. 4, pp. 4021–4026 (2001)
16. Gao, D., Xue, D.: Terminal Sliding Mode Adaptive Control for Robotic Manipulators. In: The Sixth World Congress on Intelligent Control and Automation, vol. 2, pp. 8853–8857 (2006)

Author Index